Risk, Governance and Society
Volume 15

Marion Dreyer • Ortwin Renn
Editors

Food Safety Governance

Integrating Science, Precaution and Public Involvement

Editors
Dr. Marion Dreyer
DIALOGIK
Non-Profit Institute for Communication
and Cooperation Research
Lerchenstraße 22
70176 Stuttgart
Germany
dreyer@dialogik-expert.de

Prof. Dr. Dr. h.c. Ortwin Renn
University of Stuttgart
Department for Sociology of Technology
and Environment
Seidenstraße 36
70174 Stuttgart
ortwin.renn@sowi.uni-stuttgart.de

ISBN: 978-3-540-69308-6 e-ISBN: 978-3-540-69309-3
DOI: 10.1007/978-3-540-69309-3

Library of Congress Control Number: 2008944001

Cover design: WMX Design GmbH, Heidelberg, Germany

Printed on acid-free paper

Springer is part of Springer Science+Business Media (www.springer.com)

Foreword by Robert Madelin

The fundamental question of good governance has risen steadily up the political agenda in recent years. The turn of the millennium has seen a marked shift towards a more inclusive approach to policy development. Indeed the European Commission makes strenuous efforts to modernise the way it goes about its business, with particular emphasis on forecasting and measuring the impacts of its proposals and actions. The White Paper on Governance of 2001 marked the beginning of this new era, identifying five principles which serve as key drivers – openness, accountability, participation, effectiveness, and coherence. The principles of Better Regulation are now fully enshrined in the policy making process. The consultation of stakeholders prior to forming proposals has become standard practice; and likewise impact assessment.

Turning to food safety, this is a good time to reflect on governance. The new European food safety system is fully up and running and the European Food Safety Authority (EFSA) is well established. By creating a fully independent body responsible for risk assessment, the new food safety governance clearly separates risk assessment and risk management. But the setting up of these new arrangements is not an end in itself. In the dynamic and ever changing world of food production, new challenges continue to arise. We face the constant challenge of balancing the freedom and rights of individuals, industry and organisations with the need to reduce the real and potential adverse effects of products and processes on human, animal and plant health or the environment. Finding the correct balance so that proportionate, non-discriminatory, transparent and coherent actions can be taken requires a structured decision-making process, based on scientific and other objective information within the overall framework of risk analysis.

I warmly welcome this book's valuable contribution towards the ongoing development of food safety governance, and applaud the authors for their expertise and dedication to the cause – the cause we all share of seeking to ensure the very highest standards of food safety for all European citizens.

Robert Madelin
Director General
Health and Consumers
European Commission
Brussels, March 2008

Foreword by Catherine Geslain-Lanéelle

EFSA is one of the pillars of the European system for ensuring the safety of the food chain. Our raison d'être is the separation of risk assessment from risk management, a principle underpinning the White Paper on Food Safety, to ensure maximum independence and transparency in the decisions that govern the safety of foods. We operate independently from the regulatory authorities that request our scientific advice on risks. At the same time we work together in a single institutional framework with risk managers, national risk assessment bodies and other actors to co-ordinate our efforts in the interests of effective, science-based decision-making.

EFSA has always placed a strong emphasis on its own internal governance principles. We know that to ensure confidence in the decision-making processes, the institutions protecting health need to communicate clearly and demonstrate their independence, openness and transparency. That is why our Founding Regulation stresses the independence of our scientific advice and we apply a robust set of mechanisms to safeguard it including the Declarations of Interests made by our scientific experts.

To operate transparently we develop dialogue with our many stakeholders – ranging from other scientific bodies and regulators to food producers, retailers and consumer groups. EFSA proactively seeks their input through regular stakeholder meetings, in-depth scientific discussions and online public consultations. We use our website to provide maximum accessibility to our Management Board discussions and Scientific Panel meetings. We communicate our scientific findings independently to all interested parties, again co-ordinating with national authorities and risk managers to make sure consistent messages reach the different audiences concerned and in particular European consumers.

At the same time we are working with our national counterparts to help reinforce food safety governance even further, by building co-operative European networks to gather comprehensive EU-wide data, share scientific information, carry out monitoring and reporting, and support co-ordinated responses when required to issues of common concern.

I welcome the contribution this book makes to the efforts that we, the actors involved in the food safety system, are making to constantly progress and improve our working mechanisms and to develop the overall governance framework in which we operate.

Catherine Geslain-Lanéelle
Executive Director
European Food Safety Authority (EFSA)
Parma, March 2008

Acknowledgements

This book and the *General Framework for the Precautionary and Inclusive Governance of Food Safety* that it presents and critically discusses have grown out of research undertaken within one of the subprojects (work package 5) of the research project SAFE FOODS, 'Promoting Food Safety through a New Integrated Risk Analysis Approach for Foods'. The Integrated Project SAFE FOODS has been funded by the European Commission under the 6th Framework Programme (April 2004 to June 2008) and coordinated by Dr H.A. Kuiper and Dr H.J.P. Marvin of RIKILT-Institute of Food Safety at the University of Wageningen in the Netherlands. Subproject 5 of SAFE FOODS has dealt with institutional aspects of food safety governance with a focus on ways (procedural and structural mechanisms) to improve the implementation of precaution, participation and a politics-science interface, and has been coordinated by the editors of this book. The *General Framework* and this book have been a collaborative effort of subproject 5 in which all contributors to the first part of this book were involved. We have very much appreciated this exceptionally fruitful cooperation. It has always been both greatly intellectually inspiring (with many intensive, focused discussions) *and* very pleasant (highly cooperative and reliable).

Our efforts in developing the General Framework have benefited a lot from the valuable feedback obtained from many colleagues of SAFE FOODS at the various project meetings. Further, we owe a considerable debt of gratitude to the participants in the workshops that we undertook to reflect an early version of the governance concept with the perspectives, insights and experiences of non-governmental organisations, industry actors, risk managers and risk assessors, all of whom were selected from across Europe. We acknowledge with appreciation that these knowledgeable and experienced individuals took their time to commit themselves to our concept and provide us with feedback and advice at the deliberative events. We are especially grateful to the commentators who have taken additional time and effort to compose thought-provoking and constructive written statements about the revised governance framework that appear in the second part of the book. In addition, special thanks go to those who have helped in the production of the book manuscript, including Charlotte Reule-Giles who polished the English language of those who are no native speakers, and Jörg Hilpert who supported us in bringing the whole manuscript up to a publishable standard. Last but not least, we would like to express our special gratitude to the European Commission for financing the SAFE FOODS project and making the whole endeavour possible.

Stuttgart, 31 July 2008,
Marion Dreyer
Ortwin Renn

Contents

Contributors

David Atkins is joint Head of the Chief Scientist Team at the UK's Food Standards Agency. He has a Bsc in biological chemistry from Manchester University and a Ph.D. in environmental chemistry.

Sue Davies is Chief Policy Adviser at Which?, the UK consumer organisation, working mainly on food policy issues, and member of the EFSA Management Board; she was the first Chair of the EFSA Stakeholder Consultative Platform and is also the EU Co-chair of the Transatlantic Consumer Dialogue's Food Group.

Marion Dreyer is Deputy Scientific Director of DIALOGIK (non-profit institute for communication and cooperation research), Stuttgart, Germany, and holds a Ph.D. in Social and Political Sciences from the European University Institute, Florence, Italy.

Adrian Ely is a research fellow at SPRU-Science and Technology Policy Research, University of Sussex, and a member of the ESRC STEPS Centre. Since 2000 his research has focussed primarily on the regulation of agricultural and food biotechnology in Africa, China, Europe and the USA.

Julie Norman is joint Head of the Chief Scientist Team at the UK's Food Standards Agency. She has a Bsc in chemistry from Exeter University and a D.Phil in physical organic chemistry.

Hubert P.J.M. Noteborn is Deputy Director of the Office for Risk Assessment, Food and Consumer Product Safety Authority (VWA), The Hague, The Netherlands, and holds a Ph.D. in Mathematics and Life Sciences from the Utrecht University, Utrecht, The Netherlands.

Ruth Rawling is Vice President of Corporate Affairs EMEA for Cargill, and Chair of both Amcham EU's Agrofood Committee and of COCERAL's (Comité du Commerce des céréales, aliments du bétail, oléagineux) Food and Feed Safety Committee, Brussels, Belgium.

Ortwin Renn is Professor and Chair of Environmental Sociology at the University of Stuttgart, Germany, Director of both DIALOGIK and of the 'Interdisciplinary Research Unit on Risk Governance and Sustainable Technology Development' (ZIRN) which is part of the 'International Center for Cultural and Technological Studies' (IZKT) of Stuttgart University.

Andy Stirling is Professorial Fellow and Director of Science at SPRU-Science and Technology Policy Research, University of Sussex, and co-director of the ESRC STEPS Centre at the University of Sussex, United Kingdom.

Ellen Vos is professor of European Union Law at the Faculty of Law, Maastricht University, The Netherlands, and holds a Ph.D. in Law from the European University Institute, Florence, Italy.

Frank Wendler is Lecturer for EU Studies at the Institute for Political Science of the Unversity of Frankfurt/Main, Germany, and holds a Ph.D. in Political Science from the University of Göttingen, Germany.

Acronyms and Abbreviations

AFSSA	Food Safety Agency (France)
ALARP	As low as reasonably practicable
BfR	Federal Institute for Risk Assessment (Germany)
BMELV	Ministry of Food, Agriculture and Consumer Protection (Germany)
BSE	Bovine Spongiform Encephalopathy
Bt	*Bacillus thuringiensis*
BVL	Federal Office of Consumer Protection and Food Safety (Germany)
CAC	Codex Alimentarius Commission
CEC	Commission of the European Communities
CGB	Commission for Genetic Engineering (France)
CIAA	Confederation of the Food and Drink Industries of the EU
COCERAL	European Grain Traders Association
DALYs	Disability-adjusted life years
DG SANCO	Directorate General for Health and Consumer Protection (European Commission)
EC	European Community
ECJ	European Court of Justice
ECSC	European Coal and Steel Community
EFSA	European Food Safety Authority
EGE	European Group on Ethics in Science and New Technologies
EP	European Parliament
EPA	Environmental Protection Agency (US)
EU	European Union
FAO	Food and Agriculture Organisation of the United Nations
FEFAC	European Feed Manufacturers' Federation
FSA	Food Standards Agency (UK)

General Framework	General Framework for the Precautionary and Inclusive Governance of Food Safety in Europe
GFL	General Food Law
GM	Genetically modified
GMO	Genetically modified organisms
HACCP	Hazard analysis and critical control points
HSE	Health and Safety Executive (UK)
IAC	Interface Advisory Committee
IRGC	International Risk Governance Council (Geneva)
ISC	Interface Steering Committee
MS	Member State (EU)
NGOs	Non-Governmental organisations
NOAEL	No observable adverse effect level
NRC	National Research Council (US)
OECD	Organisation for Economic Co-operation and Development
QSAR	Quantitative Structure-Activity Relationships
QUALYs	Quality-Adjusted Life Years
PLH	Panel of Plant Health (EFSA)
PRAPeR	Pesticide risk assessment Peer Review (EU)
rBST	Recombinant Bovine Growth Hormone
SACs	Scientific Advisory Committees (FSA, UK)
SAFE FOODS	'Promoting food safety through a new integrated risk analysis approach for foods' (EU Integrated Project, 6th Framework Programme)
SCFCAH	Standing Committee on the Food Chain and Animal Health (European Commission)
SPS Agreement	World Trade Organisation agreement on the application of sanitary and phytosanitary measures
STOA	European Parliament's scientific and technological options assessment unit
STS	Science and technology studies
ToR	Terms of reference
TSE	Transmissable Spongiform Encephalopathies
vCJD	Variant Creutzfeldt-Jakob Disease
VWA	Food and Consumer Product Safety Authority (The Netherlands)
WHO	World Health Organisation
WP5	Work Package 5 (of the EU Integrated Project SAFE FOODS)
WTO	World Trade Organisation

List of Figures

List of Tables

Part I
A General Framework for the Precautionary and Inclusive Governance of Food Safety

Introduction

M. Dreyer and O. Renn

Since the mid-1990s, following a series of food-related scares and debates, with Bovine Spongiform Encephalopathy (BSE) and genetically modified (GM) foods as the most prominent issues, food safety institutions in Europe have been facing growing demands for a more effective, efficient and, at the same time, balanced and fair regulatory process that is also characterised by more transparent and participatory decision-making procedures. These demands have been motivated by concerns that powerful economic and political interests would be advanced at the expense of consumer interests – with increasing pressures resulting from broader developments such as economic globalisation, societal fragmentation, and trade liberalisation. These recent developments tend to place time constraints on all actors, create undue opportunities for special interest groups to influence the decision-making process and exert pressure on the scientific assessment process to provide results that reflect popular sentiments or easy solutions to complex problems. Food substances, products, or production techniques were sometimes represented as "certainly safe" while in fact uncertainties were denied or ignored, scientific studies not properly acknowledged, public concerns not taken seriously and, as recent food scares have revealed, even public health protection compromised.

These demands and worries have been interpreted by academics and policy makers alike as manifestations of serious legitimacy problems. By the late 1990s the prevailing diagnosis in European policy circles was that the level of public trust in both food safety and food safety institutions had seriously declined and that institutional frameworks needed improving in order to restore public trust and social legitimacy. At the level of the European Union (EU) and also in a number of EU-Member States food safety institutions were subjected to review and reform. The core of the reforms at EU-level is the allocation of responsibilities for risk assessment and risk management to separate institutions destined foremost to ensure the independence of scientific analysis and advice. This division of responsibilities is codified in the new European Parliament and Council Regulation 178/2002, widely known and referred to as the "General Food Law" (hereinafter GFL). Another prominent feature of reform of EU food safety regulation are efforts to advance the democratic quality throughout the risk regulation process, mostly by improving transparency with a focus on increased documentation and by providing more opportunities for eliciting stakeholder viewpoints.

M. Dreyer and O. Renn (eds.), *Food Safety Governance*,
DOI: 10.1007/978-3-540-69309-3_1,

This book ties in with these recent reforms and provides suggestions for carrying them forward through a set of additional procedural innovations and institutional improvements. We refer to the reforms that we recommend as the *General Framework for the Precautionary and Inclusive Governance of Food Safety in Europe* (in short the "General Framework"). This governance framework pertains to a set of challenges which we consider worthy of more attention and being in need of further advancement. These governance challenges include:

- The demarcation and coordination between assessment and management of food safety threats;
- The handling of scientific uncertainty;
- The increase of transparency during the entire food safety governance process;
- The involvement of a diversity of social groups and the wider public into the governance process;
- The handling of highly controversial food safety issues.

These issues are all addressed – at least to some extent – by the recent EU-level reforms. These reforms, though significant, do not fully address prominent concerns and criticism. The results of the empirical research which was carried out to inform the development of the General Framework (these results will be described in more detail in Chap. 1) suggest that both the issues and the recent reforms that have an impact on them continue being subjects of debate and controversy. The question of how to organise the relationship between scientific expertise and political decision-making in the governance of food risks, which was placed high on the European policy agenda mainly due to the BSE crisis, is still not sufficiently solved in the view of many practitioners and concerned or interested observers. It is precisely through the full organisational separation of risk assessment responsibilities (which lie with the European Food Safety Authority, EFSA, located in Parma) from risk management responsibilities (which lie with the EU institutions, i.e. European Commission, European Parliament and the Council/Member States) that it has increasingly become articulate that scientific activities cannot be performed in complete isolation and in a political vacuum. The famous National Research Council's "Red Book" has already pointed out a central and well-founded criticism of "full organizational separation" which states that "simply separating risk assessment from the regulatory agencies would not separate science from policy" (NRC 1983: 139). How then to account for the inherent interlinkage between the scientific and the political aspects of food safety governance *without* compromising the generally agreed functional differentiation between activities aimed at "understanding" risks and activities aimed at "acting" on risks? And how to create transparency on the way in which this complex and close relationship is dealt with?

Official representations in EU food safety regulation increasingly express commitment to a more systematic recognition, consideration and communication of the scientific uncertainties that may be involved in the assessment of risk. At the centre of a more systematic approach to dealing with the challenge of scientific uncertainty lies the application of the precautionary principle, formally established by the GFL as a general principle of food law. However, there are a number of questions for its application in food safety governance which are subject to fierce debates. In particular, there is the question over whether precaution is applicable to assessment at all, or whether it is simply an approach to risk management. Alternatively, if precaution is applicable in the assessment stage, what is then the precise nature of the relationship between precautionary approaches to assessment and established practices based on conventional risk assessment? Furthermore, how could more clarity be produced over the *triggering* of the precautionary principle and provisions established to ensure that the principle is applied in a more consistent, predictable and non-arbitrary manner?

In the past four years there have been growing efforts to involve stakeholders in both management and assessment of food safety threats. Still, there is ongoing intense debate over the question of how to involve efficiently and legitimately both corporate and civil society actors in food safety regulation, especially in conditions of social controversy. This question gained prominence through both the BSE crisis and the persistent debate on GM crops and food. Currently, it is increasingly being discussed in relation to topics such as the use of animal cloning for food production, the methods of characterising genotoxic substances in food, and a broad range of potential applications that rely on nanotechnologies. The need for reconsideration of stakeholder involvement in the regulatory process in face of these "old" and emerging issues is widely acknowledged. At the same time the question over how to feed the perspectives of a wide diversity of social groups and also of the wider public systematically into the regulatory process, without an overkill of participatory procedures that would abuse the scarce resources of both the responsible institutions and those "involved", becomes more important. "Stakeholder fatigue" seems to develop into a buzzword in academic and stakeholder circles. Moreover, the consultation of stakeholders through the assessment authority, EFSA, remains a disputed issue. At the core of this debate is the question of how to ensure that this does not compromise the safeguarding of assessment against "inappropriate" non-scientific influences.

The governance framework which will be presented in the *first part* of the book suggests a set of procedures and structures that the General Framework envisages to improve the dealing with these particularly challenging governance issues in a transparent and politically accountable manner. These innovations are able to further implement the principles of good governance enshrined in the General Food Law and the agenda on governance in the European Union.

The General Framework is not the result of research work carried out in academic isolation. A first version of the governance framework had been subjected to a *systematic feedback and review process* in form of a series of four workshops with key actors in the field of food safety governance at which this early concept was presented and discussed. This process of stakeholder engagement was undertaken through the autumn of 2006 and involved, successively, industry representatives (Haigerloch/Germany, Castle of Haigerloch), representatives of non-governmental organisations (NGOs) (London, British Academy), risk managers (Brussels, Fondation Universitaire) and risk assessors (Brussels, Fondation Universitaire). At these workshops important insights were gained into the practicability and political and social viability of the governance framework. The review and feedback process was completed on 11 May 2007 when the refined and elaborated governance framework was presented at a final workshop (Brussels, Fondation Universitaire). The objective of this Presentation Workshop was to reflect the amended version with the views of those who had contributed to the feedback process hitherto and with the perspectives, insights and experiences of a wider audience in order to complement the final concept. The *second part* of the book is dedicated to input and commentaries by key actors in food safety governance. It will point out how the deliberative feedback events shaped the development and final design of the governance framework, and provide four commentaries on the final concept as presented in the first part of the book from individuals with wide experience of food safety governance who participated at the May Presentation Workshop.

In the remainder of this introduction, the content of the chapters of the two parts of the volume will be sketched.

Part 1: *The General Framework for the Precautionary and Inclusive Governance of Food Safety in Europe*

Chapter 1 will elaborate on the *challenges* that European food safety governance is facing at present and point out the *policy imperatives*. This will be done with reference to the current legal and policy framework, and to viewpoints and experiences of key actors of food safety governance elicited in our empirical research in order to inform the development and design of the General Framework. The chapter will set out that any innovative food safety governance framework will need to address, clarify and carry forward the main elements in current EU law and policy on the governance of food safety (most notably the General Food Law), the implementation of precaution (notably the European Commission's Communication on the Precautionary Principle, CEC 2000a, which was broadly endorsed by EU Heads of Government in a European Council Resolution at Nice in December 2000[1]), its relationship with overarching principles of good governance (as discussed in the Commission's White Paper on European Governance, CEC 2001a) and with established international frameworks (notably World Trade Organisation and Codex Alimentarius). Further, the chapter will introduce the *key conceptual ideas* upon which the governance framework builds in order to respond to the major policy imperatives.

Chapter 2 will first set out *historical precedents* of the proposed General Framework. This discussion will focus on three models (drawing on a conceptual distinction introduced by Millstone et al. 2004): the simplistic "technocratic" model, wherein objective science is seen to directly inform policy making; the "decisionist" model, which corresponds closely to that illustrated by the National Research Council's Red Book and recognises that policy making requires inputs other than science in order to inform decisions; and the "transparent" model, that recognises the formulation of "social framing assumptions" based on socio-economic and political considerations. It will be pointed out that, in line with the "transparent" model, the book advocates a governance concept that aims to build transparency in

[1] See Presidency Conclusions, Nice European Council Meeting 7, 8 and 9 December 2000.

decision making around European food safety by explicitly recognising the function of the *framing* step. Then, the chapter will provide an *outline* of the overall architecture of the General Framework (moreover inspired by the conceptual work of the International Risk Governance Council, IRGC) and its individual components. The major components are the governance stages of *framing, assessment, evaluation* and *management*, and the two cross-cutting activities of *participation* and *communication* which constitute integral parts of all four stages.

Chapter 3 will focus particular attention on the more detailed structure of the processes of *framing*. It will set out that "risk assessment policy" (in the terminology adopted by Codex Alimentarius), the importance of which in influencing decisions around food safety has been highlighted by various recent studies, falls within these processes. Against the background of empirical insights, it will be recommended that this policy should be understood as a task to be undertaken *jointly* by assessors and managers, in a fashion that is transparent to and takes account of inputs from a wide range of stakeholders. Three major stages of framing will be identified. First, there is *review*, the ongoing process of adapting and improving the arrangements for food safety governance within the EU to respond to the global contexts in which they are situated. Second, there is *referral*, the process of referring a specific case (be it a new food product, production method, industrial process, or commercial practice) to the European Food Safety Authority (EFSA) for screening and later for assessment. Third, there is the process of setting detailed *terms of reference*, including information on the most appropriate assessment approaches for a specific case, upon which EFSA should act and issue a scientific opinion.

Chapter 4 is dedicated to those activities carried out solely by assessors, largely EFSA, focussing on the work of EFSA under the proposed governance framework. It will present the four different approaches to assessment that the General Framework distinguishes. These are *presumption of prevention, precautionary assessment, concern assessment*, and *conventional risk assessment*. In the proposed framework, efficient and effective allocation to the different assessment processes is achieved by means of a series of explicit criteria, against which each food safety threat in question is examined, and which are developed during the process of *review* at the framing stage. The chapter will describe the more detailed structure of this activity of *screening* – which is based on a distinction between the attributes of *seriousness, uncertainty*, and *ambiguity* – and treat the actual use of the screening criteria. In particular, this chapter will establish a basis for understanding the modalities for the implementation of the *precautionary principle* in assessment, and the detailed implications for the role of conventional as well as more elaborate forms of assessment.

Chapter 5 will deal with the more detailed processes of *evaluation* and *management*. It will point out that in the General Framework the *tolerability/acceptability judgement* at the evaluation stage is informed, but not determined by the results of the assessment process. Evaluation implies that the insights of the assessment exercise are deliberated in consideration of *wider social and economic factors*, and that the necessary *trade-offs* are made between threats, benefits, and other relevant impact categories taking account of multiple perspectives. The results of this *balancing process* inform a decision on the necessity of a management process and the selection of appropriate management measures. The chapter will treat the series of steps involved in the decision-making process on management measures, and distinguish, in analogy to assessment, four approaches to management – *prevention*, *precaution-based*, *concern-based*, and *risk-based*. Each of these approaches lends itself to a set of suitable management measures. While there is no automatic correlation in the allocation of assessment and management approaches, there is a preliminary assumption that the appropriate assessment approach is subsequently pursued during the phase of management. The chapter will set out that evaluation – in the same way as framing – should be understood as a joint task of assessors and managers requiring inputs from a variety of stakeholders.

Chapter 6 will deal with the legal and institutional conditions and requirements to implement the proposed framework in current EU food safety governance and make suggestions for *institutional integration and adaptation*. The core of these suggestions will be to establish an innovative *food safety interface structure*. The chapter will discuss three different options for the institutional design of the food safety interface and identify the establishment of an *Internet Forum* in combination with the setting up of an *Interface Advisory Committee* (composed of representatives of the Commission, EFSA, and key stakeholder groups) as the most appropriate option to improve the inclusiveness, transparency and coherence of procedures at the assessment/management interface. It will view the capacities of EFSA to conduct the various tasks of screening and assessment and recommend that a *Screening Unit* and a *Panel on Concern Assessment* would be created in order to improve these capacities. The chapter will then discuss compatibility of these proposals and the overall architecture of the proposed governance framework with general principles of European law, requirements established through case law of the European Court of Justice, and international agreements especially in the framework of the World Trade Organisation (WTO).

While all previous chapters already touched on the topic of participation, Chapter 7 will provide a condensed presentation of the participatory design of the governance of food safety as envisioned by the General Framework. It will start out with featuring the special value that is assigned to the *interface structures* (the Internet Forum and the Interface Committee in two variants) as formal mechanisms for putting the idea of *inclusive governance* into practice. These structures will be characterised as *permanent deliberation and consultation platforms* aimed at facilitating the coordination between assessment and management, and at addressing the concerns of a diversity of social groups throughout the governance process. Then the chapter will deal with the question of how to specify whether it could be required to resort to more extensive participation in a given case, i.e. to select *additional* participatory processes (extending beyond Internet Forum und Interface Committee). As a second major provision for a more structured approach to participation the chapter will offer a default assumption that under the conditions of high levels of scientific uncertainty and/or socio-political ambiguity, a higher degree of participation is required. The question of what follows the requirements for extended participation will be discussed with regard to each of the four major governance stages.

Chapter 8 will present the approach to communication on food safety issues as envisioned by the General Framework. It will provide an outline of the *evolution* of "risk communication" practices and set out that the General Framework's approach is inspired by the rationale of the current (third) development phase. This phase stresses a two-way communication process in which it is not only the members of the public who are expected to engage in a *social learning process*, but the assessment actors and food safety managers as well. The chapter will specify the major *functions* of food safety communication, and point out the *requirements* for communication for each stage of the governance cycle. Then, it will present an overview of tools which can be assigned to communicative and dialogue-driven procedures, distinguishing three basic types of tools (*information-based, dialogue-based, participation based*). The chapter will explain the need for systematic *evaluation* of communication efforts in order to assess their effectiveness and discuss a number of ways to perform an evaluation. It will present a set of *principles* of good food safety communication practice and explain the ways in which communicative procedures, if adhering to these principles, can contribute to enlightenment, confidence-building, and improved coping with food safety threats by influencing behaviour.

In order to demonstrate how the General Framework introduced in the preceding chapters could be *implemented*, Chap. 9 will work through the *case* of placing on the market for consumption as food of Bacillus thuringiensis (Bt) Cry1Ab transgenic *Zea mays*. Bt maize is among the first generation of genetically modified foods that were submitted for regulatory appraisal within the European Union, and several events have received food safety clearance from EFSA. The chapter will mention aspects of each of these historical cases. However, in order to demonstrate the proposed governance framework as clearly as possible, the *hypothetical* case study will assume current levels of scientific knowledge as if a new Cry1Ab event were submitted for human food use under the contemporary legal framework. The case study will run through each of the governance stages outlined in Chaps. 3–5 individually. The chapter will not make prescriptive judgements regarding decisions that the respective institutions should make (e.g. around terms of reference, screening criteria or assessment outcomes). Instead, it will explain the *mechanisms* through which each of these stages would be executed, suggest *possible results* at each of these junctures and explain the *potential consequences* in terms of subsequent stages in the governance framework.

Chapter 10 will provide a summary of the *key features* of the proposed General Framework and specify the way in which these features relate to established principles of food safety governance as enshrined in the General Food Law and high profile general agendas around the governance of European Union institutions. The summary will highlight provisions of the General Framework to avoid overburdening the food safety governance system and overexploiting scarce financial and staff resources for making decisions. These provisions are key to making the framework *practical*. They are based on a careful distinction between different aspects and contexts of food safety each demanding at least partly-distinct technical methodologies, deliberative processes and institutional configurations. The chapter will set out that the main contribution of the General Framework has been to scope out a minimal and straightforward way in which these complex demands might be reconciled.

Part 2: *Input and Commentaries by Key Actors in Food Safety Governance*

The first chapter of this part of the book, Chap. 11, will provide a synopsis of the input of the *workshop-based feedback and review process* into the development of the General Framework. It will highlight major viewpoints gathered throughout the series of workshops with key actors in the field of food safety governance and delineate the way in which an earlier version of the governance framework was revised in response to this feedback. The need for reconsideration of the suggestions for institutional reform will be pointed out as the *main lesson* that could be learnt from the deliberative exercises. The chapter will set out how the revised suggestions for institutional reform, as presented in detail in Chap. 6 of this book, seek to respond to the following two issues raised across the consulted actor groups: first, how to achieve a high degree of *inclusiveness* in the food safety interface activities; and second, how to design structural devices that promise to promote continuity, transparency and accountability in the activities of screening, setting the terms of reference and evaluation without rendering the governance system overly complex and eventually inert.

Chapter 12, then, is composed of *four commentaries* to the revised governance framework as presented in the first part of the book which we requested from four individuals who are all professionally involved with aspects of handling food safety threats. We asked them to critically review the General Framework from their specific professional perspective. The four sections of Chap. 12 provide commentaries in the perspective of:

- Risk management by Dr. David Atkins and Dr. Julie Norman, Head of the Chief Scientist Team, Food Standards Agency (FSA), London, United Kingdom;
- Risk assessment by Dr. Hubert P.J.M. Noteborn, Deputy Director of the Office for Risk Assessment, Food and Consumer Product Safety Authority (VWA), The Hague, The Netherlands;
- A consumers' association by Sue Davies, Chief Policy Advisor, Which?, London, United Kingdom;
- Industry by Ruth Rawling, Vice President of Corporate Affairs EMEA for Cargill, and Chair of both Amcham EU's Agrofood Committee and of COCERAL's Food and Feed Safety Committee, Brussels, Belgium.

The variety of aspects that the commentaries deal with includes those, mentioned above, of inclusiveness and the delicate balance between transparency (especially of the interface activities setting the terms of reference and evaluation) and manageability of processes and structures. These aspects are among those substantive issues which will continue to be of great importance and challenge in our research on the governance of risk, uncertainty, and ambiguity. The thoughtful and stimulating reflections of the four critical reviews will be of great help in our future attempt to further address these issues.

Chapter 1
The Need for Change

A. Ely, A. Stirling, M. Dreyer, O. Renn, E. Vos, and F. Wendler

1.1 Fundamental Challenges

The governance of food safety presents a formidable series of challenges, both in general and, more specifically, within the context of the European Union.[1] The purpose of this chapter is to outline and explore some of these challenges, bringing into focus the conceptual ideas upon which we may build in order to address them. The existing conditions that necessitate change in food safety governance arrangements within the EU will be discussed and related to potential procedural and institutional responses. As such, this chapter introduces and defines the terms used to describe the various stages in the governance process, as well as some of the specific problems encountered during each of these activities. These concepts will be further expanded upon in subsequent chapters describing a general framework for food safety governance within the European Union that can address the challenges discussed here.

1.1.1 Conceptualising Stages in the Governance Process

For the purpose of our analysis, as in discussions of other 'technological risk' issues, the governance process is understood to include, but also to extend beyond, the three conventionally recognised elements of *risk analysis* – risk assessment, risk management, and risk communication.[2] *Governance* thus includes matters of institutional design, technical methodology, administrative consultation, legislative procedure and political accountability on the part of public bodies, and social or

---

[1] Many of these challenges are set out in Regulation (EC) No 178/2002 (*OJ* 2002, L31/1) as amended by Regulation (EC) No 1642/2003 (*OJ* 2003, L 245/4), hereinafter referred to as the *General Food Law* (GFL), and also referred to in other European Commission documentation, such as the White Paper on European Governance (CEC 2001a), and the Precautionary Principle (CEC 2000a).

[2] See National Research Council (NRC) (1996), and Codex Alimentarius Commission (CAC) (2005), GFL.

M. Dreyer and O. Renn (eds.), *Food Safety Governance*,
DOI: 10.1007/978-3-540-69309-3_2, © Springer-Verlag Berlin Heidelberg 2009

corporate responsibility on the part of private enterprises. But it also includes more general provision on the part of government, commercial and civil society actors for building and using scientific knowledge, for fostering innovation and technical competences, for developing and refining competitive strategies, and for promoting social and organisational learning.

Within this broad notion of governance, the General Framework outlined in Chap. 2 moves beyond the elements of risk analysis to account for the processes through which policy problems are identified as such, and the institutional and political influences that shape the ways in which these problems are perceived, conceptualised and prioritised by policy makers. This element of the governance process is here termed *framing*. Encompassing activities such as the identification of the scientific inputs required to inform policy, framing sets the terms of reference for the next stage in the governance process: *assessment*. Assessment subsumes, with other methods which will be described in more detail in Chaps. 2 and 4, the conventional procedures of 'risk assessment' as variously defined. Through gathering information on technical and socio-economic risks and benefits, as well as on the concerns of stakeholders and citizens, assessment informs, substantiates and justifies governance decisions, policies and wider institutional practices and commitments. The framework proposed in this book suggests two further stages that contribute to the goals of food safety governance. Based on the outputs of the assessment, an *evaluation* exercise is undertaken. This exercise summarises the information gathered during the assessment phase and involves deliberation around divergent values associated with the threats under consideration. Following the evaluation exercise, intervention measures are identified, assessed, and selected in a process of *management*. This process also includes the implementation of such measures and their follow-up through monitoring existing threats and horizon scanning for emerging threats.

1.1.2 Precaution as a Response to Lack of Scientific Certainty

One of the most significant challenges for risk governance relates to current and highly topical debates over the application of the *precautionary principle*. Variously defined in a multitude of different instruments, this embodies the central injunction that lack of scientific certainty should not be used as a reason to delay appropriate action.[3] It is in this form that precaution has become a guiding principle of EU policy making[4] and is recognised by the European Court of Justice and the Commission of the European Communities (CEC 2000a) to be a general principle of European law. Yet this raises a number of profound questions for its

[3] As expressed, for instance, in the classic definition at Principle 15 in the 1992 Rio Declaration on Environment and Development.

[4] See Article 174(2) of the EU Treaty.

application in food safety governance. In particular, there is a question over whether precaution is applicable to assessment at all, or whether it is simply an approach to risk management (ESTO 2000; Harremoes et al. 2001; CEC). Alternatively, if precaution is applicable in the assessment as well as in the management stages of food safety governance, then there follow an entire series of more detailed queries over the precise nature of the relationship between precautionary approaches to assessment and established practices based on conventional risk assessment. One central feature of this relationship stems from the formal scientific definition of the condition of '*risk*' (Knight 1921) itself.

Over many decades of intensive academic and policy activity, the term *risk*, properly speaking, refers to a situation in which it is possible confidently to quantify both the magnitudes of and the probabilities for a defined range of outcomes (such as forms or degrees of harm in food safety).[5] Indeed, it is this central reliance on probabilities that is a key diagnostic feature of conventional approaches to risk assessment. Variants of these probabilistic 'risk-based' methods offer sophisticated responses to different forms of *complexity* in social, technological and natural systems (IRGC 2005). In the food safety realm, for example, probabilistic techniques might be applied to the characterisation of risks from a chemical additive with well-characterised toxicity and substantial long-term data on consumption levels. In a more complex case, probabilistic modelling might be used to investigate the potential synergistic activities between this chemical additive and a natural toxin existing in a traditional food product in which consumption patterns are well characterised). However defined, the precautionary principle addresses a set of more intractable circumstances – going beyond complexity – under which various forms of 'incertitude' render such quantification incomplete or problematic (World Trade Organisation 2004; Public Health Reports 2002).

These more intractable circumstances can take three main forms, which are illustrated in Fig. 1.1 below. The first is referred to in the strict definition of the state of '*uncertainty*', under which the possible outcomes are clear, but it is difficult to quantify probabilities (Knight 1921; Keynes 1921). As demonstrated in the figure, an example might be the potential for cancer associated with a novel carcinogen.

[6]The second is the condition of '*ambiguity*', where the problem lies not with probabilities, but in agreeing the appropriate values, priorities, assumptions, or boundaries that apply in defining the possible outcomes (summarised in Stirling 2003). Questions around the tolerability of a new form of battery husbandry with animal welfare implications could produce a condition of ambiguity. Third, a condition of '*ignorance*' exists where neither probabilities nor outcomes may be fully or confidently characterised. In this latter case, where 'we don't know what we don't know', we are seeking to mitigate our exposure to surprise (Shackle 1955; Loasby 1976). At the level of the UK in the early 1980s, when BSE was first

[5]This is also a key element in the seminal formal understanding of risk assessment promulgated in the NRC's 1983 'Red Book'.

[6]The general scheme here is taken from Stirling (1999). Examples have been added for the purposes of this particular exercise.

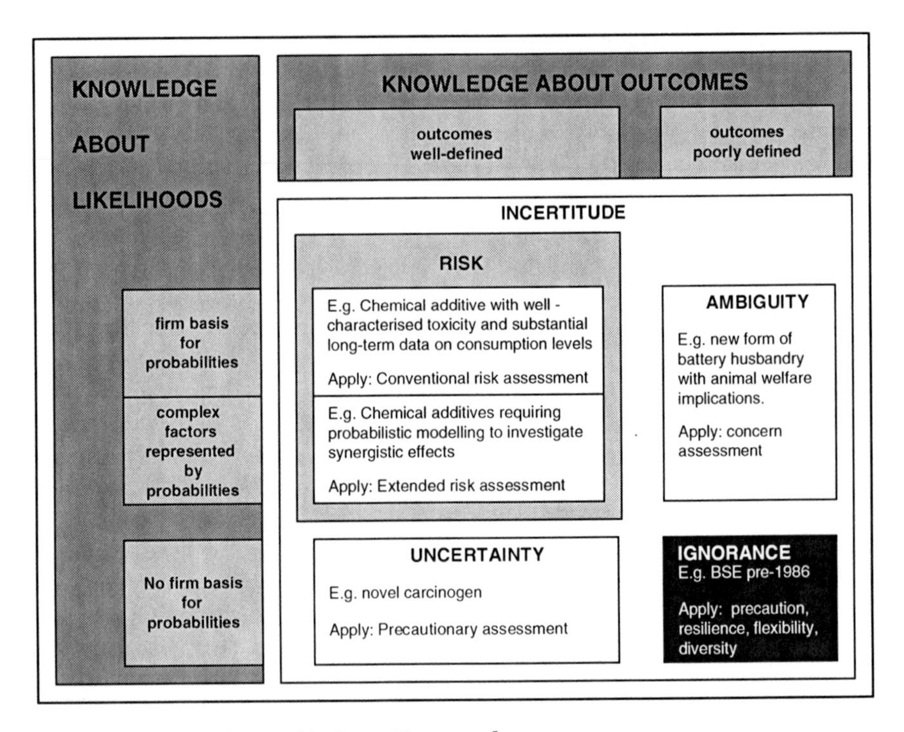

Fig. 1.1 Risk, uncertainty, ambiguity and ignorance[7]

appearing in cattle, ignorance existed as to the number of associated potential human deaths from variant Creutzfeldt-Jakob disease (vCJD) and the probability that these deaths would emerge. Various forms of conventional risk assessment remain applicable under conditions of complexity. But uncertainty, ambiguity, and ignorance are, by definition, states of knowledge under which conventional probability-based risk assessment is quite simply inapplicable (Stirling 1999). In such cases we look towards resilience, flexibility, and diversity in agri-food systems in order to allow effective responses to areas of ignorance once they have been identified. Where conventional risk assessment leaves residual uncertainties unaddressed, these must be addressed by other complementary methods. It is in recognition of this challenge that we find the basis for reconciling conventional risk assessment and precaution in terms of their complementarity.

In short, the direct implication of the precautionary principle for assessment is to highlight the conditions under which it would be appropriate to apply what may be described as more *comprehensive* approaches to assessment (GFL, Art. 7). These are noted in Fig. 1.1 above and will be discussed in much greater detail in Chaps. 2 and 4.

[7] The general scheme here is taken from Stirling (1999). Examples have been added for the purposes of this particular exercise.

These fundamental challenges to the stage of assessment raise some important implications for the current, conventional practice in the governance of food safety of opting by default for the application of conventional risk assessment. Unconstrained reliance on established risk assessment methods can sometimes seem to reflect a rather narrow and complacent view of uncertainty and an optimistic or expedient view of the depth and form of knowledge that is necessary in assessment (ESTO 2000). In governance terms, this can present problems of coherence, effectiveness, accountability, and participation. On the other hand, recourse to more comprehensive but demanding 'precautionary' approaches to assessment can bring its own problems. To some, precaution can appear unduly pessimistic about the quality of the available knowledge. In particular, there can be a lack of clarity over the 'triggering' of precaution, and the consequent procedures may seem fuzzy, onerous, erratic, or disproportionate in their effects (Miller & Conko 2001). These can raise different challenges of timeliness, proportionality, predictability, and consistency – as well as coherence in the articulation of conventional risk assessment and precaution. Chapter 4 will provide a detailed examination of these issues.

1.1.3 Resultant Questions

In a field like food safety with its public profile and global importance, these challenges introduce very high political, economic and institutional stakes. Each side of the conventional risk assessment/precaution contrast is thus characterised, in different ways, by various actors for contending purposes. Whatever the details in specific instances, the general effect is to compound the prevailing state of confusion, polarisation, and conflict over the appropriate approaches to assessment. Yet, despite the complexities, the central challenges seem quite clear. In short, any governance framework for food safety must address the following five questions:

(a) How can governance address elements of *risk*, *complexity*, *uncertainty*, *ambiguity*, and *ignorance* in ways that are open, coherent, effective, accountable, and participatory?
(b) In particular, how can we articulate relatively narrow forms of conventional *risk assessment* with more comprehensive forms of assessment suggested by the *precautionary principle*, in a fashion that is coherent, operational, proportionate, and consistent with wider governance principles?
(c) What are the appropriate roles for different specialist disciplines, technical procedures, institutional designs, and modes of engagement under different forms of assessment and at different stages of the governance process, and how should these relate to each other?
(d) How can *framing*, *assessment* and *evaluation* reflect different forms of knowledge, contested political–economic interests and socio-cultural values in a balanced fashion, such as to provide those who *manage* a given threat with the

broad-based knowledge necessary to yield feasible, timely, proportionate and consistent – as well as socially legitimate and robust – governance outcomes?

(e) How do the proposals regarding safety governance outlined here relate to existing procedures and institutional arrangements in Member States and at the EU level? To what extent can the proposed framework be accommodated by current arrangements which are centred solely on conventional ideas of risk assessment and risk management?

Each of these questions will be addressed in forthcoming chapters. In order to provide further context for their treatment, the next section will outline the policy imperatives for improved food safety governance.

1.2 Policy Imperatives

In order to set out the policy imperatives, this section will first highlight some of the major recent institutional re-arrangements and efforts into procedural reform in food safety regulation and sketch the legal and policy basis on which these changes and reform efforts build. In a second step, it will point out certain issues that emerge as essential to the task of changing food safety governance to the better. It will do so by reference to the policy imperatives identified in the legal and policy documents. In addition it will refer to policy imperatives which key stakeholders in the field emphasise on the basis of some years of experience since the changes have been introduced.

The exposition draws on the results of two empirical activities. First, it takes up the insights gained in a comparative study of institutional re-arrangements in food safety regulation that have taken place over the past decade in Europe. This study includes the EU-level and five European countries: Hungary, Sweden, France, the United Kingdom, and Germany.[8] While the results on the EU case are of overriding importance for the purpose of the present exercise, insights gained from the country studies will be set forth where appropriate. Pertinent are empirical insights in relation to the challenges implied in the division of institutional responsibilities for risk assessment and risk management which characterise the EU food safety system as well as the French and German systems. The second source of empirical information on which the following sections draw is a series of workshops with key actors in the field of food safety governance at which a draft version of the governance framework presented in this book was put forward for discussion. The feedback events were conducted through the autumn of 2006 and involved, successively,

[8] The results of this study, also carried out within subproject 5 of the SAFE FOODS project, are presented in Vos and Wendler (2006a). At EU-level and in each of the five countries, semi-structured interviews were carried out with risk assessors, risk managers, and key stakeholders. A total of 13 interviews were conducted at EU-level, 12 in Sweden, 16 in the United Kingdom, 23 in Hungary, 24 in France, and 25 in Germany.

industry representatives, representatives of NGOs, risk managers, and risk assessors, all of whom were selected from across Europe.[9]

1.2.1 Recent Institutional and Procedural Reforms in Food Safety Governance

Over the past decade, food safety regulation at EU-level and in several EU Member States represents a highly dynamic policy field, subjecting institutions to considerable pressure to demonstrate competence, credibility and fairness in the handling of risk problems. This pressure has resulted first of all from the experience of a gradual but substantial withdrawal of public trust in both food and those responsible for food safety, following a series of food-related scares, most notoriously the BSE crisis of the mid 1990s. Since then, food safety institutions in Europe have faced a crisis of social legitimacy. Empirical research has shown that this crisis has triggered noticeable institutional responses designed to restore public trust and social legitimacy.

There are at least three responses that stand out: *First*, there is the use of mechanisms designed to assure a stricter separation of the risk assessment function from political decision making. Providing the public with an independent and disinterested expert view about the magnitude of a risk through scientific analysis, and then explaining and justifying the regulatory actions that are based on these scientific assessments, has come to be recognised as a major step towards more transparency, accountability and, in particular, trustworthiness. In terms of loss of trust, the remedy resorted to in this approach is the trust-generating power of what is represented as '*independent risk assessment*'.[10] Safeguarding scientific analysis against distortion by inappropriate policy influences and considerations is intended to re-establish and assure the credibility of risk assessment activities and results on which risk management decisions are to be based. This approach is especially pronounced at EU-level and in those countries, including Germany and France,[11] where responsibility for the functions of risk assessment and risk management has been allocated to different institutions.

---

[9]For each of the workshops a summary report was produced and circulated to the workshop participants after the event to ensure accuracy and provide the opportunity for further feedback.

[10]While official rhetoric often evokes the idea of '*science only*' in this respect, scholars in the field of science and technology policy have persuasively argued that this model, even in theory, is misleading: The specific approach of a particular risk assessment, including, e.g. the selection of impacts to assess, the disciplinary perspectives to shed light on these impacts, and the choice of more or less conservative safety factors, does inevitably involve non-scientific considerations and value judgements, be they explicit or implicit; cp. Millstone (2000: 118); Millstone and van Zwanenberg (2002: 603); Jensen and Sandøe (2002); see also the NRC's 'Red Book' of 1983 which argues that the description of risk assessment as a strictly scientific undertaking was a misconstruction (NRC: 150).

[11]In France, however, trust-building appears to rank behind improvement of effectiveness as a rationale for separation (Dreyer, Renn, Borkhart, & Ortleb 2006: 19).

This, as the European Parliament's Scientific Technology Options Assessment (STOA) 2000 study points out, clearly contrasts with the practice prior to the 'BSE-turning point', when both EU institutions and EU Member States were neither systematically differentiating between activities of risk assessment and risk management, nor did they structurally separate organisational or institutional responsibilities (Trichopoulou et al. 2000: 67). It was normal for the responsibility for assessment and management to be handled by a single institution, for those responsible for risk management to be closely involved in preparing and deciding scientific characterisations of risks, and for scientific advisors to be expected to provide specific advice on particular policy issues (*Ibid.*). Since that turning point, however, the appropriateness of this approach has been challenged in the scientific as well as policy communities. The BSE crisis was interpreted as a result, at least partly, of a regulatory regime marked by a non-transparent intermingling of the roles of assessment and management, and of scientific and non-scientific considerations. The Committee of Inquiry into BSE, set up by the European Parliament, in its *Medina Ortega* report deemed a blurred relationship between science and policy and a lack of transparency to have been major shortcomings of the EC's policy – in the years before 1996 – as well as of the British approach. It concluded that the EU institutions had given precedence to national interests of agriculture and industry at the expense of public health protection (Vos & Wendler 2006b: 69).

Suspected of abetting partiality and obscurity in dealing with food risk issues, the traditional approach of rather seamless scientific and political activities became a subject of intense debate, scrutiny, and reform. It is the *primary feature* of the current institutional framework of EU food safety regulation that the responsibilities for assessment and management are divided between institutions, with the newly established European Food Safety Authority (EFSA) being located in Parma and the European Commission in Brussels.

A policy of reassurance linked to a partial treatment of scientific information has been described as one of the principal shortcomings in the UK's policy-making on BSE until the mid 1990s. It was pursued, despite a lack of certainty that BSE posed no risk to humans, it undermined precaution, and it eventually produced a legitimisation crisis when in March 1996 UK government ministers announced that BSE had most likely been transmitted to humans (van Zwanenberg & Millstone 2001). It seems reasonable to assume that the growing attention to and communication about *scientific uncertainties* at the EU-level is at least in some part a response to the UK's critical experiences in terms of a 'lesson learnt'.

Official EU statements increasingly declare scientific uncertainties to be an important subject of assessment, a component of transparency and public communication, and a matter of accountability in their own right. For example, EFSA has set up a Working Group to develop a framework for a guidance document dealing with transparency in risk assessment, including the way in which adequate information on the strengths and weaknesses of the data used could and should be provided (Vos & Wendler 2006b: 106). In a 2005 discussion paper, the European Commission's Health and Consumer Protection Directorate General (DG SANCO) moreover critically notes that public debate would tend to over-sell science as a source of certainty. In order to achieve clearer risk perceptions and a better integration of risk into EU policy debate, according to the paper, it is of great importance that the limits on scientific certainty are more accurately understood, and that the responsible authorities are able to highlight and communicate scientific uncertainties (DG SANCO 2005).

The more careful consideration of scientific uncertainties can be understood as a *second* resource employed to address a situation of low trust and legitimacy.[12] Just as the provisions for enhancing the independence of risk assessment, it can be described as a *results-based* legitimacy mechanism.

The EU as well as the UK have also resorted to reforms designed to hold up the *procedural legitimacy*[13] of food safety governance by incorporating democratic norms in the risk analysis process.[14] Advancement of the *democratic quality* of the governance process forms the *third* major response to the situation of "*contested governance*" (Ansell & Vogel 2006).[15] It is formulated on the DG SANCO's website as follows:

Transparency of legislation and effective public consultation are essential elements of building this greater [consumer] confidence.[16]

There are three major modes by which this purpose was expected to be served in food safety regulation:

- Making the risk analysis process, including risk assessment, more transparent through wider public documentation (including the publication of EFSA's opinions on the Authority's website);
- Providing more opportunities for the consultation of economic and civil society actors in relation to both assessment activities (with EFSA's Stakeholder Consultative Platform taking a prominent position) and management activities

---

[12] Attention to and communication of scientific uncertainties seem to be rarely directly represented as trust-building measures. However, this point of emphasis usually forms part of official representations of the new approach to food safety governance, which typically include more or less specific references to the trust issue.

[13] The exposition adopts here the argumentation by Grace Skogstad, who suggests in her analysis of GMO regulation in the EU that, "all strategies to render policies acceptable by virtue of democratising the procedures by which they are arrived at, can be viewed as input-oriented legitimation" (Skogstad 2003: 324). While the "test of appropriateness" under output, or results-based, legitimation standards was the perceived merit of policy outcomes, this test under input, or procedure-oriented, legitimation standards was the conformity of decision-making procedures with democratic norms of public participation and control (Skogstad: 324–325).

[14] To a lesser extent the same holds true for France and Germany, which have also declared the (re-)establishment of consumer confidence as one objective of their revised food safety policy (Dreyer, Renn, Borkhart, & Ortleb 2006: 51–58).

[15] The editors of this book refer to the situation of "both sudden and pervasive loss of trust and legitimacy and an uphill battle to restore it" (Ansell & Vogel 2006: 20) as "contested governance" and argue that European food safety regulation over the past decade exemplified such a case.

[16] http://ec.europa.eu/food/food/foodlaw/principles/index_en.htm. Accessed 30 May 2008.

(with the Advisory Group on the Food Chain and Animal and Plant Health taking a prominent position);
- Offering more comprehensible and process-oriented information on risk to the public at large, specifically addressing major consumer concerns.

In short, the shift to procedurally-based legitimacy as a supplement to results-based legitimacy includes efforts to provide public access to documentation of both the outcomes *and* the procedures of both risk management *and* risk assessment, to consult with commercial and civil society actors on a more *regular* basis and in a more *open* manner (which contrasts with informal and confidential 'behind-closed-doors' consultations), and to provide the public at large with more targeted information (Dreyer, Renn, Borkhart, & Ortleb 2006: 30–45).

1.2.2 Governance Aspects in Need of Further Improvement

At EU-level, the most specific and authoritative codification of current structures and practices including the institutional re-arrangements and reform efforts set out above is provided in the European Parliament and Council Regulation 178/2002 on general principles and requirements of food law and setting up the European Food Safety Authority of 2002, better known and throughout this book referred to as the 'General Food Law' (GFL).[17] Grounded in a wider regulatory literature (NRC 1983/1996; EPA 1997; CAC 2005), this rests on three key pillars. The first pillar is the application of principles of independence, objectivity and transparency in *risk analysis* (as defined in Sect. 1.1.1), the second pillar is the application of the *precautionary principle* in the face of scientific uncertainty, and the third pillar is the resort to *public consultation.*

Public consultation directly relates to *participation* as one of the five normative principles of *good governance* that the European Commission has identified in its White Paper on European Governance. It requires governance institutions actively to engage with other social groups, from the conception of strategic options right through to the implementation of decisions. The four other principles are openness, accountability, effectiveness, and coherence (CEC 2001a), all of them directly applicable to the good governance of food safety. According to the Commission the principle of *openness* entails clear, accessible communication of the nature and rationale for decisions and other governance outcomes. *Accountability* involves clarity over the nature of the reasoning and the allocation of responsibility in legislative and executive processes. *Effectiveness* relates to timeliness, delivering what is needed on the basis of clear objectives, and an impact evaluation. It includes issues of subsidiarity and proportionality in decision outcomes. *Coherence* concerns the degree of consistency that can be achieved by complex institutional frameworks in addressing even more complex technical, social, and natural systems.

[17] See discussion in Vos and Wendler (2006b).

It is important to note that the revised European food safety governance system embedded in this legal and policy framework is an *evolving* system. Many specifications of the recent reforms are still very much developing. It is an inherent part of this embryonic stadium of change that the challenges of putting the reforms into practice in an effective and politically and socially acceptable manner are becoming increasingly visible. The following sections address some of these challenges. It will be argued that in order to further implement the principles of food safety governance enshrined in the General Food Law and the agenda on governance in the European Union several aspects deserve more attention and need further improvement.

1.2.2.1 Reconsideration of the Relationship Between Risk Assessment and Risk Management

As set out above, the division of responsibilities for risk assessment and risk management between institutions rests on one of the major pillars of the General Food Law which is the application of the principles of independence, objectivity and transparency in risk analysis. This substantial institutional re-arrangement is intended to ensure primarily the political independence of the risk assessment authority and a disinterested scientific description of food safety issues. While separate responsibilities are generally seen as a welcome development, political decision makers, scientific experts, and economic and civil society actors increasingly realise that the institutional and geographical segregation of risk assessment creates new challenges in terms of organising the *relationship with risk management.*

In the first couple of years after EFSA's establishment much of the official rhetoric tended to evoke the idea of assessors and managers doing their jobs in strict separation and sequence. Various interviewees and also several participants at the workshops with key actors in food safety governance stressed, however, that this concept has never presented practical reality in which interaction occurs and is deemed necessary. There, obviously, exist tensions between public legitimisation needs (insulating science from policy) and practical action requirements. Interviews with policy actors and expert advisors at EU-level, in France, and also in Germany indicated that the experience with the new institutional divide has increasingly brought to light that problems might arise if the need for interaction is not accounted for at specific points in the risk governance process (Dreyer, Renn, Borkhart, & Ortleb 2006: 27–30). The two main actors at EU-level, EFSA's Scientific Committee (EFSA 2006a: 9) and the Commission's DG SANCO, have recently explicitly recognised the need for an "efficient and transparent mechanism of interaction" between risk assessment and risk management (DG SANCO 2005).[18]

[18]This corresponds with the Codex Working Principles for Risk Analysis which emphasize that risk analysis is an iterative process and interaction between risk managers and risks assessors essential for practical application (CAC 2005, Art. 9, 102). The NRC's 'Red Book' of 1983 is emphatic on that: "The importance of distinguishing between risk assessment and risk management does not imply that they should be isolated from each other; in practice they interact, and communication in both directions is desirable and should not be disrupted" (NRC: 6).

Interaction is deemed particularly relevant at the start of the risk governance process when a problem needs to be defined and the questions and tasks for the risk assessors need to be delineated. The interviewed Commission officials emphasised the necessity to be present during meetings of EFSA's panels in order to explain their needs, to better understand the reflections of the scientists, to change the terms of reference if deemed necessary by EFSA, and also to make sure that a panel is not stepping in risk management issues (Vos & Wendler 2006b: 119). Also in France, the stage of *framing* the issue and of setting the terms of reference has been identified as a critical issue in terms of interaction. The French food safety agency, AFSSA, has addressed this issue by introducing *quality procedures in referral handling*. These include training, ad hoc rather than systematic, of ministry personnel by assessors to assist those in the ministry in phrasing referrals properly (Mays, Jahnich, & Poumadere 2006: 231–233).

A second interaction issue, brought to light by the comparative study, relates to the power of the risk assessment authorities to publish autonomously. From the interviews, it could be concluded that EU and also French and German risk managers have increasingly recognised the need for co-operation with the assessment authorities with regard to *communicating food risks to the public*. They expressed a preference for a buffer period before the publication of risk assessment opinions and related press announcements during which they could read and consider the opinion, and, if required, come back to the assessment authority for clarification or discussion of particularly important management issues. This would enable them to reflect on the management implications before being dragged into the limelight by the media and to provide both the media and the public with informed and coordinated responses.[19] EFSA and the Commission, as well as the German Federal Institute for Risk Assessment (BfR) and the German Ministry of Food, Agriculture and Consumer Protection (BMELV), have responded to this need by agreeing informally on timely information and consultation (Vos & Wendler 2006b: 93; Böschen, Dressel, Schneider, Viehöver, & Wastian 2005: 23; Dressel at al. 2006: 302).

A third critical issue in terms of interaction was highlighted by German interviewees in particular. From the side of management it was described as a special challenge to tune expert evaluative advice along the lines of risks being 'relatively low' or 'relatively high' within the wider appreciation of political, economic and social conditions and requirements on which risk management decisions are based. To address this challenge of *coordinating evaluative judgements* would require improved interaction and communication between the BfR and the risk management authorities (Böschen et al. 2005: 28). Along similar lines, from the side of risk assessment an interest was expressed in establishing, in co-operation with the Federal Office of Consumer Protection and Food Safety (BVL, the main German risk management authority) 'Best Practices in Evaluation', which would define who – the BfR, the BVL, or both – should be given the task of performing evaluative judgments at the interface between assessment and management. Such a practice

[19] For France see Mays, Jahnich, and Poumadere (2005: 65–66); Mays et al. (2006: 282).

code could enable managers to implement similar or equivalent measures in dealing with similar risks, thus enhancing consistency in decision making on risk (Böschen et al. 2005: 39). At the risk assessors' workshop it was underlined that the existence of different cultures of risk assessment in the EU and different national perspectives of what constitutes an acceptable risk, would render *systematic and transparent* evaluation, performed jointly by assessors and managers, both a necessity and a major challenge (Dreyer, Renn, & Borkhart 2007).

It was generally felt by EU-level risk assessors and risk managers whose views were elicited in the empirical research (and also by national policy makers and scientific experts) that there is still room for improving interaction, especially with regard to the aspects listed above. During interviews in the EU-level study it was suggested for example that opening up the interaction between EFSA and the Commission on the drafting of the terms of references could allow stakeholders to provide knowledge and comments. Most of the participants at the risk assessors' and risk managers' workshops underlined the need to promote and facilitate communication and co-ordination in these respects. In current practice, the interaction between EFSA and the Commission occurs mainly in an informal or semi-formal manner and is not very transparent and systematic.[20] Still, several of the workshop participants were sceptical of formalising interaction through permanent units or committees. They worried that this could end up in further complicating an already highly convoluted governance system.[21]

1.2.2.2 Application of the Precautionary Principle in the Face of Scientific Uncertainty

It was mentioned above, that official representations of EU food safety regulation increasingly express commitment to a more systematic recognition and communication of the scientific uncertainties that may be involved in the assessment of risks. Much more than in the past, the task of scientific expert advisors is seen as including both providing information about what is known and about what is *not* known. At the centre of a more systematic approach to dealing with the challenge of scientific uncertainty (as defined in Sect. 1.1.2) lies the application of the *precautionary*

[20] In contrast with previous practice, where informal and pragmatic interaction was taken for granted, interaction is today more focused on and subjected to restriction and scrutiny. Provisions for the involvement of risk managers in the assessment process are one example of this. The respective Article of the General Food Law (28 (8)) stipulates that, if invited to do so, representatives of the Commission may assist the discussion process for the purpose of clarification of information, but they should not attempt to influence the debate. The specific unit within DG SANCO that deals with the relations with EFSA (formerly unit 5) shall fulfil a 'watchdog' function in this respect and prevent Commission officials from overstepping the role of an observer who may supply information on request.

[21] This concern was expressed most strongly at the workshop with risk managers (Vos & Wendler 2006c).

principle, the second major pillar on which the General Food Law rests. In codifying and defining the precautionary principle with particular reference to food safety, the Law directly addresses the contentious nature of the relationship between risk assessment and precaution. Drawing on concepts that are discussed in Sect. 1.1.2, the Law characterises the application of the precautionary principle in the following terms (Art. 7 (1,2))[22]:

> 1. In specific circumstances where, following an assessment of available information, the possibility of harmful effects on health is identified but *scientific uncertainty* persists, *provisional* risk management measures necessary to ensure the *high level of health protection* chosen in the Community may be adopted, pending further scientific information for a *more comprehensive* risk assessment.
> 2. Measures adopted on the basis of paragraph 1 shall be proportionate and no more restrictive of trade than is required to achieve the *high level of health protection* chosen in the Community, regard being had to *technical and economic feasibility and other factors* regarded as legitimate in the matter under consideration. The measures shall be reviewed within a reasonable period of time, depending on the nature of the risk to life or health identified and the type of scientific information needed to clarify the *scientific uncertainty* and to conduct a *more comprehensive* risk assessment. [present authors' emphasis]

In short, through its references to both more comprehensive risk assessment and provisional risk management measures under conditions of persistent uncertainty, the General Food Law acknowledges that the precautionary principle is of direct and important relevance to the assessment, as well as to the management, of food safety. Although little analysis is provided of the detailed rationale, and no examples are given fully to substantiate the concept of "*more comprehensive risk assessment*", the injunction to greater comprehensiveness clearly reflects an understanding of the circumscribed status of conventional risk assessment as an approach to promote a broader understanding in assessment.

The empirical findings indicate that while precaution is generally acknowledged as a major EU policy-making principle the concept continues to be contested in the actual regulation of risk which holds true also for the regulation of food risks. In particular, there is a lack of clarity over the 'triggering' of the precautionary principle and a related scepticism over the possibility of applying the principle in a consistent, predictable and non-arbitrary manner. The nature and extent of scientific uncertainty or evidence of the possibility of a serious risk required to justify a precautionary approach remains an open question (Vos & Wendler 2006b: 112). Another question which is deemed important, but unsettled, concerns the way in which the precautionary principle should and could be used in accordance with the principle of proportionality when deciding on management measures (Mays et al. 2006: 252).

The information gained from the interviews with decision makers and scientific advisory experts suggests that in current practice the interpretation and application of the precautionary principle varies across countries and authorities, and appears

[22] See discussion in Vos and Wendler (2006b: 112–114).

highly contingent on the respective regulatory framework, on individual cases, and on the respective case assessors and managers. Both at EU- and Member State levels, the approach to identifying, characterising, and communicating scientific uncertainties and handling them on the basis of the precautionary principle is *ad hoc* and *case-specific*, rather than systematic and based on concrete guidelines. This may be at least partly due to the rather under-specified reasoning and implications of the discussion in the General Food Law of the relationship between risk assessment and precaution.

1.2.2.3 Opening Up the Governance Process Through Public Participation

Public consultation is the third major pillar on which the General Food Law rests. It is represented as a response to the circumstance that:

> food safety and the protection of consumer interests are of increasing concern to the general public, non-governmental organisations, professional associations, international trading partners and trade organisation.[23]

The Law stipulates that, with the exception of urgent matters, there shall be "open and transparent public consultation, directly or through representative bodies, during the preparation, evaluation and revision of food law" (Art. 9). Furthermore, it specifies that EFSA shall develop "effective contacts with consumer representatives, producer representatives, processors, and any other interested parties" in the course of risk assessment (Art. 42). The Law is also specific about the participation component in risk management, which is defined as being about "weighing policy alternatives in consultation with interested parties" (Art. 3 (12)). This is in line with the concept of risk communication advocated by the Commission's White Paper on Food Safety which defines it as an interactive and involving dialogue with and feedback from stakeholders (CEC 2000b).

Up to now, one of the most notable changes to the traditional practice of involving interested and affected parties has been the fact that the risk assessment phase is being opened up to some degree to consultation. This new practice is not accepted unquestioningly. The findings of the empirical study show that the inclusion of stakeholders in the course of risk assessment is still very much disputed and has an exploratory character. By no means were all interviewees convinced about the necessity of having interested parties involved in an activity that should be governed by data gathering and analysis, and safeguarded against inappropriate non-scientific influences, but nor did they have clear ideas about appropriate ways to do justice to this legal imperative (Vos & Wendler 2006b: 124). According to various Commission officials who were interviewed in the empirical study, a viable option to greater public involvement as regards risk

[23]DG SANCO's website, http://ec.europa.eu/food/food/foodlaw/principles/index_en.htm. Accessed 30 May 2008.

assessment would be to consult stakeholders more regularly at the moment of drafting the terms of reference and after presentation of the assessment report (Vos, Ni Ghiollarnath, &Wendler 2005: 131).

Another change from the *status quo ante* in current consultation practice is represented by the greater importance being attached to the representation of consumer interests. At EU-level it is institutionalised in the Advisory Group on the Food Chain and Animal and Plant Health and the Stakeholder Consultative Platform set up by EFSA. EFSA also provides for a formal representation of consumer interests at the management level, in the Authority's Management Board. These provisions are generally welcomed by representatives of non-governmental organisations (NGOs). Various participants at the NGO workshop pointed, however, to the continuing challenges faced by NGOs around unequal power relations and access to resources between different actors in food safety governance in which informal contacts behind closed doors continue to be of high importance. In particular, the possibility that governance questions would be framed by the powerful corporate sector means, in their view, that it is important to have formal NGO involvement already at the early stage of the governance process when the problem is being defined and the terms of reference set, and certainly at those stages at which action needs and ways of action are being deliberated and concluded (Ely & Stirling 2006).

At present, public consultation is mostly organised as stakeholder consultation, giving in particular the bigger and more prominent organisations a voice within the framework of the Advisory Group on the Food Chain and Animal and Plant Health and the Stakeholder Consultative Platform.[24] Several NGO representatives challenged this practice stressing the need to recognise and respect the greater diversity of voices, perspectives, and values that are usually involved in food safety issues; at the same time they underlined the scarce resources of smaller NGOs or citizens to invest on a regular basis in the regulation of food safety (Ely & Stirling 2006). In a similar line of argument with regard to incorporating different views in society it was noted at the risk assessors' workshop that decision making at the management stage would need to be informed by knowledge about risk perceptions, otherwise it was more likely to erode public trust (Dreyer, Renn, & Borkhart 2007).

1.3 Practical Aims of the Present Exercise

Based on the policy imperatives identified in current legal and policy documents and highlighted by interviewees in the empirical studies and by key food safety governance actors during the series of workshops described above, certain issues emerge as fundamental to the task of improving food safety governance across the European Union. It is these issues that will form the normative basis for the General Framework to be introduced in subsequent chapters.

[24]To be sure, the Commission also organises regularly public consultations on several topics where everyone is invited to give comments.

In addition to those questions (a)–(e) outlined in Sect. 1.1.3 above, a number of further questions remain to be addressed in any attempt to develop a truly integrated governance framework that will address the requirements of the General Food Law, the White Paper on Governance, the multiple forms of less tractable incertitude outlined above and the other issues raised by stakeholders. In particular:

(f) How can *framing* be organised so as to engage stakeholders and the public, and to allow different perspectives and priorities to be addressed in policy formulation, in a way that addresses ambiguity and uncertainty?
(g) How can interactions and communications between the European Commission, EFSA and stakeholders be improved so that *assessments* are framed and *evaluations* concluded in an effective, open and transparent manner?
(h) Within the activity of assessment: what is the operational definition of "*persistent scientific uncertainty*" as defined in Article 7 of the General Food Law and by what practical means can it be characterised in the process of assessment?
(i) Which are the key operational features of "*more comprehensive risk assessment*" and how do they relate to current conventional and alternative available procedures?
(j) How can we decide what constitutes an appropriately "*high level of health protection*", and how exactly does this relate to "*technical and economic feasibility and other factors*"?
(k) How can we ensure that the *principle of proportionality* is upheld in a procedurally consistent manner under different situations of persistent uncertainty?
(l) How can the objectives of *openness* and *participation* be addressed in an effective and proportional way throughout the governance process, especially as regards assessment, evaluation and management?

Together with the earlier questions (a)–(e) – set out in Sect. 1.1.3 – it is these issues that must be addressed by the present candidate design for an integrated general framework for the governance of food safety. Part I of this book responds to each of these through the following chapters.

Chapter 2
Overview of the General Framework

A. Ely, A. Stirling, M. Dreyer, O. Renn, E. Vos, and F. Wendler

2.1 Historical Precedents

Frameworks for food safety governance have evolved through a variety of forms since the mid-late twentieth century, and it is useful to reflect on these developments prior to introducing the General Framework adopted in this book. The simplistic *technocratic* model, wherein objective science is seen to directly inform policy making (shown in Fig. 2.1), gave way in the late twentieth century to the less naïve *decisionist* model (shown in Fig. 2.2).[1] This model, which corresponds closely to that illustrated by the National Research Council's (NRC) "Red Book" (NRC 1983), recognised that policy making required inputs other than science in order to inform decisions, and that other legitimate factors (such as those relating to socio-political and economic objectives) needed to be taken into account in addressing risks. The Red Book in 1983 established the division between the scientific aspects (*risk assessment*) and political aspects (*risk management*) within the overall process of risk analysis. This division, and several other aspects of the Red Book model, have been adopted across a wide variety of risk management fields (Omenn 2003).[2]

Chapter 1 (Sect. 1.2.1) has discussed how recent institutional and procedural reforms in European food safety governance have continued this trend. The objective of promoting "independent risk assessment" within EFSA, as legislated for under the General Food Law, has been seen as an important condition for re-building trust in the EU regulatory process, especially following the lessons from the BSE crisis. As has been discussed in Chap. 1, however, the strict separation of risk assessment and risk management laid down in the General Food Law is in practice somewhat blurred.

[1] The distinctions between the three models outlined in Figs. 2.1–2.3 are taken from Millstone et al. (2004).

[2] It is worth bearing in mind, however, as pointed out in Chap. 1, that the view of risk assessment as a purely scientific exercise was also questioned within the "Red Book" (NRC 1983).

M. Dreyer and O. Renn (eds.), *Food Safety Governance*,
DOI: 10.1007/978-3-540-69309-3_3, © Springer-Verlag Berlin Heidelberg 2009

Fig. 2.1 The *technocratic* model (from Millstone et al. 2004)

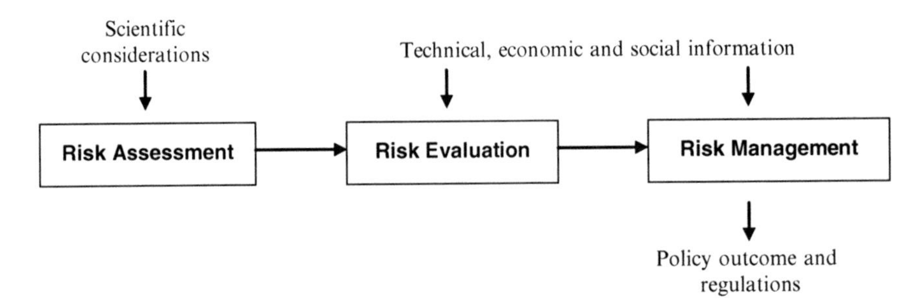

Fig. 2.2 The *decisionist* model (from Millstone et al. 2004)

Since the widespread diffusion of the risk assessment/risk management distinction, careful analyses of the role of science in policy making have increasingly pointed to the importance of "framing assumptions" in informing risk assessment. These insights have questioned the simple risk assessment/management boundary by pointing to politically informed decisions around how risk assessment should proceed. Such decisions do not necessarily determine the outcome of the scientific assessment, but may often circumscribe the scope, or at least the minimum scope, of the risk assessors' deliberations. Millstone et al. (2004) have borrowed from the terminology adopted by the Codex Alimentarius Commission to characterise these decisions as relating to "risk assessment policy". According to them, such decisions concern issues such as:

- Which kinds of impacts are deemed to be within the scope of the assessment and which were outside it,
- Which kinds of evidence to include and which to discount,
- How to interpret the available evidence,
- How to respond to uncertainties, and
- How much of different kinds of evidence would be necessary or sufficient to sustain different types of judgements (Millstone et al. 2004: 1).

Millstone et al. (2004) have thus proposed a more sophisticated model for understanding policy that recognises the formulation of *social framing assumptions* based on socio-economic and political considerations. Based on research into science-related trade disputes over beef hormones, recombinant bovine growth hormones (rBST) and GM crops they argue that policy officials are increasingly articulating a co-evolutionary model that questions the over-simplicity of the decisionist model's

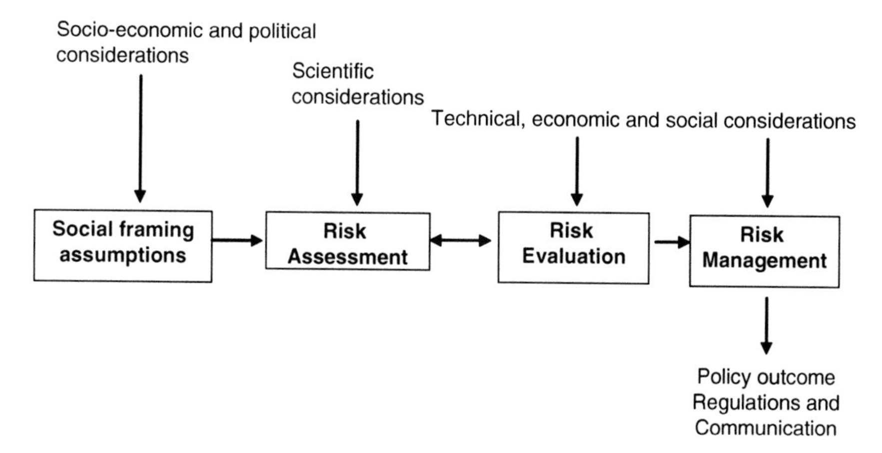

Fig. 2.3 The *transparent* model (from Millstone et al. 2004)

artificial distinction of a purely scientific up-stream risk assessment phase followed by a down-stream risk management phase. The *transparent* model (Fig. 2.3) views scientific and socio-political factors as intertwined throughout the process of policy making and communication, with reciprocal links between science and policy, and recognises the input of various actors at each stage in the process. Millstone et al. qualified their use of the word "transparent" by stressing that if current practices in policy making around food risks were conducted transparently (which largely they are not), they would be seen as operating in accordance with this model. We view *framing* as an important aspect of risk governance, advocating a governance concept that aims to build transparency in decision making around European food safety by explicitly recognising the function of this step.

While communication around risks, both with stakeholders and the public, has traditionally (at least within the technocratic model) been seen as a separate process, carried out following assessment and management, the governance approach adopted by us views *communication* as well as *engagement* with stakeholders and the public, as integrated into every stage in the process. This corresponds with the relevant texts in Articles 3 (12, 9) and 42 of the General Food Law, as previously discussed in Sect. 1.2.2. Communication and engagement within the advocated governance framework will be covered in more detail in Chaps. 7 and 8.

A simplified representation of the governance framework is illustrated in Fig. 2.4 below (the complete and detailed framework is outlined in Fig. 2.8), highlighting the successive stages of framing, assessment, evaluation, and management. Each of these stages fulfils specific roles within food safety governance, engaging stakeholders in the ways most appropriate to ensure the principles of good governance outlined in Chap. 1 (as will be covered in more detail in Chap. 7). Sections 2.3 to 2.6 of this chapter will be dedicated to outlining the function and procedural aspects of each of these stages, before they are discussed in more detail in Chaps. 3–5. The section subsequent to this stage-related outline (Sect. 2.7) will provide an overview of the major aspects of the cross-cutting activities of communication and participation.

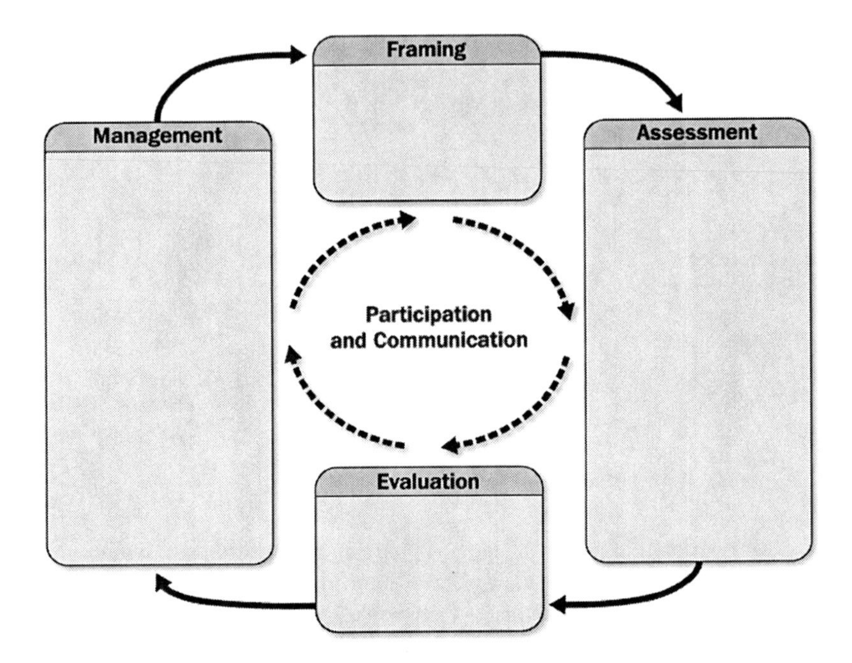

Fig. 2.4 A simplified representation of the General Framework for food safety governance

2.2 The General Framework: A Schematic Picture

In broad terms, the proposed framework includes the well-established stages of risk analysis described above, here referred to as *assessment* and *management*. Moreover, as the representation in Fig. 2.4 shows, the framework renders the established linear structure – in common with other contemporary conceptions of risk governance[3] – into an open, cyclical, iterative and interlinked process. In this respect, there is particular resonance with the broad frameworks currently emerging under the auspices of the International Risk Governance Council (IRGC 2005). Furthermore, it includes two additional governance stages: firstly, *framing* which relates to risk assessment policy (in the terminology adopted by Codex Alimentarius 2005; Millstone et al. 2004), and, secondly, *evaluation* which relates to the process of assimilating and deliberating upon the outputs of the assessment phase and considering the tolerability or acceptability of a given threat more explicitly in the governance cycle. These two stages act to promote efficient and transparent mechanisms of interaction between risk assessment and risk management. All steps of the

[3]Prime Minister's Strategy Unit/UK Cabinet Office (2002); NRC (1983); Royal Commission on Environmental Pollution (RCEP) (1998).

cycle are interlinked and involve multi-actor engagement processes that are specified in later parts of this document.

Several points are important to note at the outset, prior to the description of the advocated framework. The first is that this framework distinguishes between the *precautionary principle*, *precautionary assessment* and *prevention*. Section 1.1.2 focussed on the problem of the conditions under which the precautionary principle might be triggered by assessments of uncertainty. For the purpose of this book, and in line with the definitions given by the European Court of Justice and the General Food Law (Art. 7), we consider the precautionary principle to be a general governance principle employed in framing the overall process of framing, assessment, evaluation and management. In particular, as will be explained, precaution applies to the *screening* of food safety *threats*[4] for the properties of seriousness or uncertainty in order to determine their subsequent treatment in assessment and management. Precautionary assessment consists of a "more comprehensive" approach to assessment (as discussed in the previous chapter), adopted in cases where screening has identified a lack of scientific certainty of the kind referred to in the General Food Law. Prevention refers to the approach that is taken when a food safety threat is identified as being both serious and certain.

Secondly, it is important to note at the outset that the General Framework is primarily designed to address the regulation (including licensing) of food products, production methods, industrial processes and commercial practices. This is an extremely broad field. However, it does exclude certain important areas of regulatory activity, such as cases where developments are driven by urgent need directly to respond to particular emerging "food scares". In this latter case assessment does not necessarily begin with a particular identifiable product, process or practice. Instead, attention starts with a less readily characterisable social or public health phenomenon, for which causal relationships with particular products processes or practices may be difficult to establish. Under such conditions – though the present framework will not be irrelevant – certain additional features will be necessary, which lie beyond the scope of the present exercise.

It is further important to note that the implementation of the procedural provisions envisaged by the General Framework does not necessarily require institutional changes but could be effected through the currently existing institutional arrangements. While the governance framework outlined here introduces certain innovative elements, especially at the interface between risk assessment and risk management, it generally fits into the existing legal and institutional framework of European food safety regulation as defined by the General Food Law and other, more case-specific pieces of framework legislation (such as the regulations and directives setting out the procedures for the authorisation of GMO products) as well as the current structures and practices of food safety regulation at the European level (cp. Vos & Wendler 2006b). Against this background, it is the intention of the proposed General Framework to make recommendations especially for the

[4] For a definition of the term "*threat*" see Sect. 4.1.

improvement of *practices* and *approaches* within the conduct of risk regulation, while complying with, and further implementing the key principles of the General Food Law and other relevant legislation and case law.

The limited institutional adaptations that will be suggested would, however, *facilitate* the working of the proposed procedural reforms. In the following chapters, we refer to two major adaptations: a *Screening Unit* and a *Panel on Concern Assessment* within EFSA as part of a proposal for the improvement of the capacities of EFSA to fulfil the functions foreseen in the General Framework, and two *food safety interface institutions* to improve the inclusiveness, transparency and coherence of the setting of terms of reference and evaluation. The latter comprise an *Internet Forum* (an online function, managed by the Commission that allows open and transparent communication between the Commission, EFSA, Member States and wider stakeholder groups) and an *Interface Committee* (which may take two different forms). These limited institutional changes are discussed in detail in Chap. 6.

2.3 An Overview of Framing: Review, Referral and Terms of Reference

Framing refers largely to what may be called the "meta-level" of food safety governance, involving the whole range of processes concerning the iterative design and development of the framework conditions of regulation in the face of new learning and feedback between the various processes, both through binding rules and non-binding conventions. By explicitly including this as an element in the General Framework, it is acknowledged that the implementation of food safety governance takes place at a number of organizational, legal and discursive levels that lie outside the detailed focus of this book (for example within Codex Alimentarius or the World Trade Organisation, WTO). Framing is made up of three activities – *review* of the technical and institutional conditions relating to food safety in its broadest sense, *referral* of specific threats to EFSA for the process of screening, and the setting of *terms of reference*, upon which EFSA will base their assessment. These are represented diagrammatically in Fig. 2.5.

2.3.1 Review

Review sets the structure of the legal and institutional design with respect to responsibilities, rights, obligations, division of labour, prescribed procedures, and oversight activities. It also includes the dynamic aspect of incorporating structural changes over time and is closely related to the underlying philosophy of food safety governance. Review thus involves activities such as the development and enactment of laws and regulations (e.g. the EU's General Food Law and its regulations on genetically modified food), the generation and use of legal principles (such as the

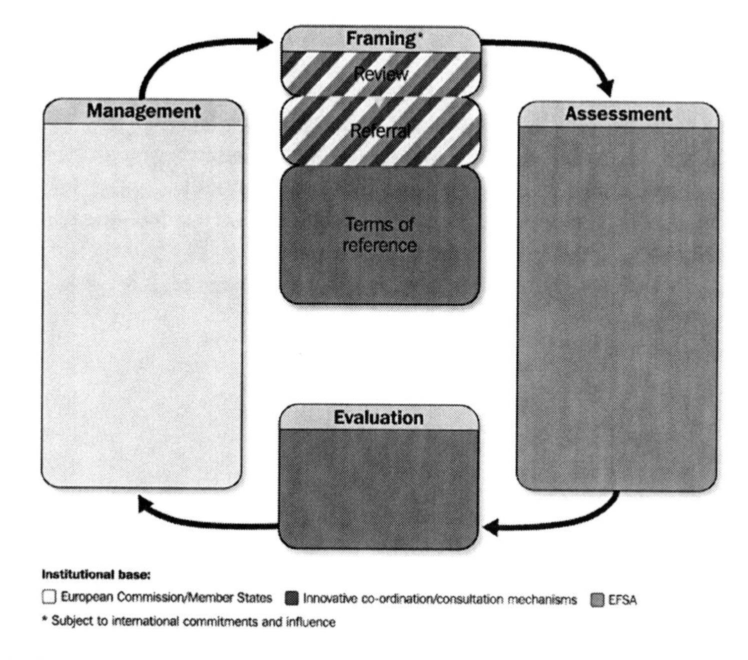

Fig. 2.5 The stages of *framing*, in relation to the rest of the governance cycle

Precautionary Principle), the determination of scientific conventions (such as statistical procedures), the establishment of predominant procedural perspectives (such as the three-step risk analysis process), and also the review of the conduct of the safety governance process as a whole. All of these activities have an impact on how the concrete design of the governance framework is spelled out and changes over time. The EU institutions are obviously highly influential in these framing activities, but also global organisations – the WTO and the Codex Alimentarius Commission, in particular – and the Member States exercise an influence.

2.3.2 *Referral*

In contrast to the structural conditions under which regulation takes place, the step referred to as *referral* focuses on the concrete processes and procedures by which food safety problems are identified, formulated, and initially referred to EFSA for screening and assessment. Referral is based upon the legally prescribed regulatory framework of a product, a production method, an industrial process, or a commercial practice. Once such a substance, process or outcome is identified as possibly being subject to regulatory actions on the basis of the general legislative provisions (on the basis of Art. 29, GFL), and has to be submitted to specific licensing, certification, or testing whether all standards are met, it is forwarded to EFSA for screening. Referral may hence be performed by applying existing laws or regulations or by initiating preliminary regulatory procedures resulting possibly in modifications of existing or even the drafting of news acts by the European institutions. The process of referral will often fall to the Commission or to Member States, however the establishment of the Internet Forum and Interface Committee will also allow the opening up of referral to a wider range of stakeholders. It is understood that in cases of self-tasking by EFSA, which are prescribed by Art. 29(1) (b) of the General Food Law, this step is omitted and the food safety governance cycle starts at the stage of screening.

2.3.3 Terms of Reference

Screening, which is carried out by EFSA and is thus described further in the following section on assessment, involves the preliminary characterisation of the threat in question in order to select the most appropriate form(s) of assessment. This assessment must be based on specific and detailed *terms of reference* (which are formulated based on an exchange of opinions by the Commission as the manager, EFSA as the assessor and the relevant stakeholders). It is during this process of setting terms of reference that residual uncertainties or data gaps in relation to a threat may be identified, or specific participatory procedures or consultations with external experts may be requested to form part of assessment. The terms of reference will be informed by the insights gained through the screening exercise in relation to what constitutes the most appropriate, efficient and proportionate form of more detailed assessment. While the drafting of the terms of reference is currently undertaken either by a specific unit of DG SANCO (in cases of a request by the Commission), or by the originator of a request, it is the intention of the proposed governance framework that this step should involve both, assessment actors and managers in conjunction with representatives of key stakeholder groups. While DG SANCO may retain the overall responsibility for the drafting process, the Internet Forum and the Interface Committee will allow these other actors the opportunity to influence and monitor the process.

2.4 An Overview of Assessment

A key element in the broader process of food safety governance lies in the assessment of risks and benefits from alternative products, processes, investments, standards, regulations, and strategies. In this document, we consistently use the broad term *assessment* (as opposed to "risk assessment" or "conventional risk assessment") to refer to the process of gathering, eliciting, synthesising and deliberating over information and perspectives that are pertinent to governance decisions. Assessment therefore subsumes, with other methods which will be described in more detail below, the conventional procedures of "risk assessment" as variously defined. It is foremost assessment that informs, substantiates and justifies governance decisions, policies and wider institutional practices and commitments. As such, assessment helps ensure coherence, inform openness and provide accountability.

2.4.1 Screening

EFSA will receive its initial mandate to assess a given food safety threat through the process of referral outlined above. The first stage in the subsequent assessment is that of *screening*, in which the most appropriate approach to assessment is identified. During the screening stage, which follows after referral, key features of the food safety threat in question are identified and pre-classified in advance of actual assessment. In the interests of openness, effectiveness and proportionality, the attributes of seriousness, uncertainty, and ambiguity are used to identify the most appropriate approach to a more detailed assessment and to help prioritise attention to different threats. This essential activity relates to established notions of *preliminary risk assessment* in discussions under the auspices of the WTO and elsewhere, which can be either quantitative or qualitative in form. Through its identification with the task of hazard identification, it is intended that this task should be undertaken by a specific unit of EFSA (a *Screening Unit*), in cooperation between the Scientific Committee or Panel and the scientific expert services. The screening process collects what is already known about the substance, process or activity (i.e. about the source of threat under consideration), characterizes the main hazard properties and suggests the appropriate assessment approach to which the threat should be submitted. The outcome of the screening process informs, as already explained above, the terms of reference.

In order to address the challenges outlined in Sect. 1.1 (surrounding uncertainty, ambiguity and ignorance), assessment within our framework includes three novel approaches in addition to the conventional risk assessment procedure. These approaches address threats which are certainly and unambiguously serious calling for a *presumption of prevention*, threats subject to scientific uncertainty calling for a *precautionary assessment*, and threats subject to socio-political ambiguity calling for a *concern assessment* (in which systematic knowledge is collected about risk perceptions by individuals and groups, socio-economic impacts and other information related to the threat source). We propose that the process of screening threats to identify which of these (or conventional risk assessment) is most appropriate should be carried out within EFSA, by individuals who have expertise not only in technical risk assessment but also in issues relating to public concerns (usually associated with the social sciences).

Based on the screening process and drawing upon stakeholder perspectives sought through the Internet Forum and Interface Committee, the terms of reference will be drafted (as mentioned above). These will include a detailed description of which approach to assessment should be followed by EFSA in order to address various aspects of the threat in question.

2.4.2 *The Four Approaches to Assessment*

The four different approaches to assessment are shown in Fig. 2.6 below. Each assessment approach is designed to gather the information necessary for making adequate and prudent governance decisions in different contexts. Where a given threat displays a number of different attributes, these different aspects may be allocated to parallel treatment by different types of assessment.

If the threats in question are certainly and unambiguously serious (illustrated by the question "serious?" in the screening stage of the diagram below), i.e. significant harm is to be expected with almost certainty, then, subject only to consideration of any overriding justification, they are assigned directly to *preventive measures*. If the threats in question are minor, and quantitative data about probabilities and magnitudes is either available or easy to produce, then they are assigned directly to *risk-based assessment.* Here there may be a presumption in favour of approval, subject

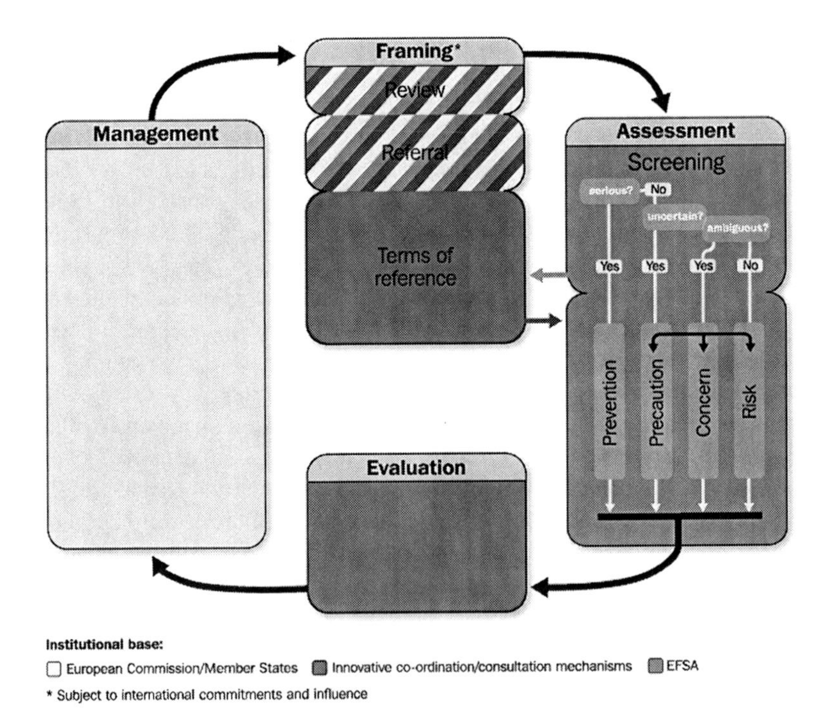

Fig. 2.6 The four approaches to *assessment*, and their relationship to *screening* and the other stages in the governance cycle

to evaluation and management considerations around the complexity and scale of the threat in question.

If screening is unable to allocate threats to straightforward preventive measures or to risk-based assessment, then more comprehensive assessment procedures are recommended. If a lack of scientific certainty has been identified in screening (illustrated by the question "uncertain?" in the same diagram), then the subsequent approach to assessment is *precautionary*. If socio-political ambiguity (illustrated by the question "ambiguous?") has been identified, then a process of *concern assessment* is adopted in subsequent assessment. Both conditions (uncertainty and ambiguity) can apply at the same time and for the same assessment candidate. In this case both approaches, i.e. the precautionary assessment approach and the concern assessment approach, need to be *combined*. Each of the four assessment approaches are discussed in more detail in Chap. 4.

2.5 An Overview of Evaluation

The step of *evaluation* which follows after the assessment stage is undertaken on the grounds of provisions of the General Food Law (Art. 3 (12)) requiring risk managers to consider "other legitimate factors" (i.e. wider societal and economic concerns) in addition to the results of the scientific risk assessment. Evaluation serves two main purposes:

- First, to reach a balanced, value-based judgment on the tolerability or acceptability of a given food safety threat, or to perform a trade-off analysis of a set of functional equivalents (of the product, process, or practice which is the threat source under consideration);
- Second, to initiate (if deemed necessary) a management process and make preliminary suggestions for the most suitable management approach.

The term *tolerable* refers to an activity that is seen as warranted on the grounds of associated benefits, yet which requires additional measures in order to reduce the threat below reasonable limits. The term *acceptable* refers to an activity where any residual threat is so low that additional measures for mitigating the threat are not seen as necessary. To draw the line between "intolerable" and "tolerable", as well as "tolerable" and "acceptable", is one of the most difficult tasks in the governance of food safety.

The tolerability or acceptability judgement is informed by the results of the assessment process but it is not determined by it. Other important considerations on wider social and economic factors may be included transparently in the balancing process. The main elements of this process are:

- The summarizing of the results of the assessment process in terms of the likely consequences for food safety or other relevant endpoints (such as environmental quality, nutrition, etc.) if no management measures were taken;

- Deliberation over these results in consideration of wider social and economic factors (e.g. benefits, societal needs, quality of life factors, sustainability, distribution of risks and benefits, social mobilization and conflict potential), legal requirements and policy imperatives;
- Weighing pros and cons and trading-off different (sometimes competing or even conflicting) preferences, interests, and values.

While assessment deals with knowledge claims (around what are the causes, and what are the effects), evaluation deals with *value claims* (around what is good, acceptable, and tolerable). Defined as a tolerability or acceptability judgement, evaluation takes up and at the same time specifies what the General Food Law refers to as the task of "weighing policy alternatives in consultation with interested parties, considering risk assessment and other legitimate factors" (Art. 3(12)). While the General Food Law determines this task as an element of risk management alongside "if need be, selecting appropriate prevention and control options" (Art. 3(12)), the General Framework, as it is presented here, refers to it as a *separate step* in the overall safety governance process *mediating* between the two stages of assessment and management.[5] Ideally, this step should, like the setting of terms of reference, involve both assessment actors and managers in conjunction with representatives of key stakeholder groups. This is best accomplished through the application of the Internet Forum in order to open up evaluation to the widest possible values base, and the Interface Committee to enable direct co-ordination between managers, assessors, and stakeholders.

2.6 An Overview of Management

As in conventional understandings of the governance of food safety, the final major stage envisaged by the General Framework is *management*. As a part of the framework presented here, it has essentially the same meaning as the definition

[5] Handling threats will inevitably be directed by evidence claims *and* normative claims. It is true that providing evidence is always contingent on existing normative axioms and social conventions. Likewise, normative positions are always enlightened by assumptions about reality (Ravetz 1999: 647–653). The fact that evidence is never value-free and that values are never void of assumptions about evidence does not compromise the need for a functional distinction between the two. For handling threats one is forced to distinguish between what is likely to be expected when selecting option X rather than option Y, on one hand, and what is more desirable or tolerable: the consequences of option X or option Y, on the other hand. It is hence highly advisable to maintain the classic distinction between evidence and values, and also to affirm that justifying claims for evidence vs. values involves different routes of legitimisation and validation. This is one of the main reasons for making an analytical distinction between assessment, evaluation and management.

given in the General Food Law (Art. 3(12)) and is, therefore, conducted by both the Commission and the Member States. Based on the output of the evaluation exercise, it is at this point that *decisions* on management measures are taken. This requires the consideration of policy choices among contending possible management measures. Such measures may include numerical limits for concentrations of substances in food items, standards for production and consumption, performance control, food preparation guidelines, monetary incentives, labels, and others. In some ways, this is analogous to the process already undertaken in assessment and evaluation. Here, however, the information is based on the positive and negative implications of a series of different regulatory interventions and not of particular threats. Depending on the context, the relevant information might best be gathered through assessment, by reference to the most relevant measures. In other cases, it will be necessary to undertake this information-gathering process at the management stage in addition – and as a complement – to the evidence gathered during assessment.

Either way, the series of steps involved in the decision-making process around management measures is as follows (cp. IRGC 2005: 40–48):

- Identification of possible measures (with special consideration of the suggestions made during the evaluation stage);
- Assessment of measures (with respect to predefined criteria);
- Evaluation of measures;
- Selection of one or more appropriate measures.

As in the assessment stage, there are various approaches to management which may be more or less appropriate in dealing with decision-making around specific measures. These broadly follow similar themes to the assessment approaches outlined in Sect. 2.4 above, but the assessment approach for a specific threat that was identified in screening does not automatically determine the most appropriate management approach. The process of evaluation, especially through eliciting value preferences around tolerability and acceptability from stakeholders, will play a large part in determining the appropriate management approach. The finer details of this process are discussed in Chap. 5 on evaluation and management.

In the broader understanding of management, this stage involves two more steps:

- Implementation of measures, and
- The monitoring of how these measures perform in practice.

Note that monitoring the outputs and effectiveness of management may lead to problems to be reframed, thus completing the food safety governance cycle. The stage of management, along with its institutional base (primarily the European Commission and Member States) and the relationship to other stages in the governance process, is illustrated in Fig. 2.7 below.

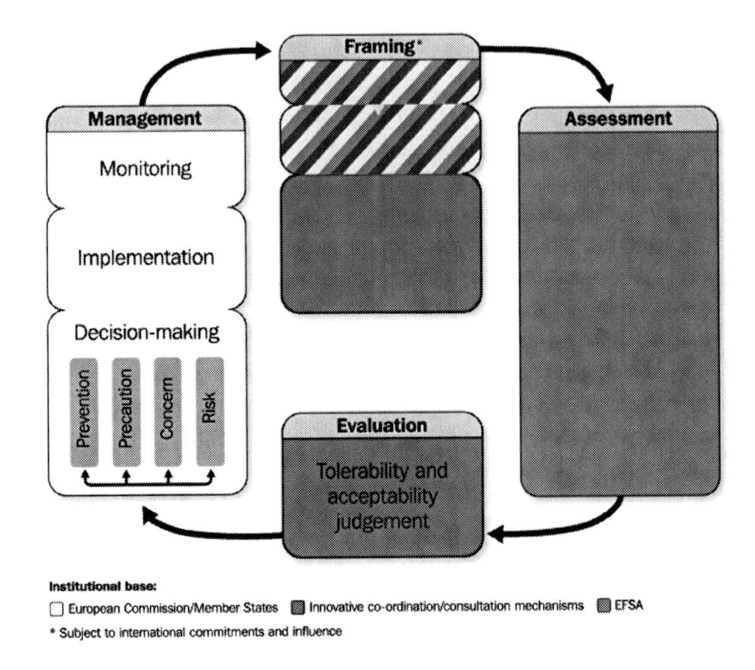

Fig. 2.7 The primary features of *management*, and their relationship with the other stages in the governance cycle

2.7 An Overview of Communication and Participation

Effective *communication* and *public involvement* are at the core of any successful activity to assess and manage food safety threats. Both tasks are placed in the middle of the food safety governance cycle (see Fig. 2.4). They constitute *integral parts* of all four stages: framing, assessment, evaluation, and management. In particular, the General Framework advocates to replace the traditional paradigm of collecting data, decision making and defending what has been decided by a new concept of an open and transparent governance process, enriched by multiple opportunities for stakeholders to feed back their knowledge and values, and a constant activity to communicate information on process as well as results to a wider public (IRGC 2005: 54).

The field of risk communication initially developed as a means of investigating how expert assessments could be communicated to the public best, so that the tension between public perceptions and expert judgement could be bridged. In the course of time, this original objective of educating the public about risks has been modified and even reversed. The professional risk community has realised that most members of the public refused to become "educated" by the experts, but rather insisted on alternative positions and risk management practices being selected by the professional community in their attempt to reduce and manage food safety threats (Leiss 1996: 85ff; Plough & Krimsky 1987).

The General Framework provides for communication about food safety threats throughout the governance cycle, from the framing of the issue to the monitoring of the management impacts. The precise form of communication needs to reflect the nature of the threats under consideration, their context and whether they arouse, or could arouse, societal concern. Communication, as advocated by the General Framework, is a means of ensuring that:

- Those who are central to framing, assessment, evaluation, or management understand what is happening, how they are to be involved, and, where appropriate, what their responsibilities are (internal communication).
- Others outside the immediate processes of framing, assessment, evaluation, or management are informed and engaged (external communication).

Although food safety communication implies a stronger role for the risk professionals to provide information to the public rather than vice versa, the governance framework, as it is proposed here, regards it as a *mutual learning process* in line with the requirements of good governance including transparency, accountability, and legitimacy. Concerns, perceptions and experiential knowledge of the targeted audience(s) should thus guide assessors and managers in their selection of topics and subjects: it is not the task of the communicators to decide what people *need* to know, but to respond to questions of what people *want* to know.[6] Communication on food safety threats requires professional performance both by food safety and communication experts. Scientists, communication specialists, and regulators are encouraged to take a much more prominent role in food safety communication, because effective communication can make a strong contribution to the success of comprehensive and responsible food safety governance.

In addition to the need for food safety communication at all stages, the General Framework provides input on all governance levels from a diversity of social groups. It promotes the idea of *inclusive governance* understood as the obligation to ensure the early and meaningful involvement of all stakeholders and, in particular, civil society (Jasanoff 1993: 123–129). Inclusive governance is based on the assumption that affected and interested parties have something to contribute to the governance process and that mutual communication and exchange of ideas, assessments and evaluations improve the final decisions, rather than impede the decision-making process or compromise the quality of scientific input and the legitimacy of legal requirements.[7] As the term governance implies, analysing and managing food safety threats cannot be confined to private companies and regulatory agencies. It rather involves a wider array of actors: political decision makers, scientists, economic players, and civil society actors.

There are two major provisions envisioned in the proposed governance framework to further improve the interaction of these actors. The first of these are the *food safety interface institutions*, the Internet Forum and the Interface Committee. They present permanent deliberation and consultation platforms to facilitate the coordination between assessment and management and to address the concerns of corporate and civil society actors throughout the governance process. The Internet Forum, our basic

[6] For an explanation of the "right-to-know" concept, see Baram (1984).

[7] Similar arguments in Webler (1999) and Renn (2004).

recommendation for creating a food safety interface structure, should act as a site for the dissemination of information associated with every stage in the governance process in order to promote the governance principles of openness and accountability. It should be designed in such a way as to facilitate proportionate deliberation between the core institutions of food safety governance with stakeholders and citizens. The modalities for ensuring effective, but proportionate, deliberation through this route are outlined in Chap. 6. It should provide an outlet for framing (e.g. referring to the appropriate European and international frameworks at issue). It can act as a dissemination and deliberation mode for the outputs of EFSA's engagement activities, particularly the Stakeholder Consultative Platform (formalized membership), annual colloquia (by invite/expressions of interest), technical meetings (by invite/expressions of interest), and science conferences and scientific colloquia (by invite). In addition, many of EFSA's current practices for public consultations and requests for data should be made more easily available to risk managers and stakeholders through hosting on the Internet Forum. These include various activities linked to assessment, such as EFSA's Pesticide Risk Assessment Peer Review (PRAPeR, 40-day consultation for new pesticide draft assessment reports), public consultations on genetically modified organisms (GMOs), additives, products and substances in animal feed, biological hazards, science committee consultations, requests for data on scientific issues, corporate events, and "Porte Aperte" (engagement with the public in the Parma region).[8] The forum would also act as a site where the Commission's consultations and decisions could be relayed transparently to the European public, allowing accountable demonstration of effectiveness and coherence in decision making. We propose to combine the Internet Forum with an Interface Committee (which is discussed in two variants with different degrees of formalisation and scope of mandate in Chap. 6). This Committee would bear responsibilities for the two interface activities of setting the terms of reference and evaluation, and composed of representatives of the Commission, EFSA and key stakeholder groups.

Specific food safety cases may require that participation through the Internet Forum and the Interface Committee is complemented by additional participatory instruments. As a second major provision to improve further the involvement of corporate and civil society groups into the governance process, the General Framework offers a *default assumption* that under the conditions of high levels of scientific uncertainty and/or socio-political ambiguity, the use of further participatory processes is required. Chapter 7 provides an outline of the implications for participation of such challenging cases in relation to each of the four governance stages.

2.8 Summary

As has been stressed throughout the present part of this book, it is important that food safety governance can adapt to the identification of new uncertainties or ambiguities within an open, iterative governance cycle. In certain cases, this may require

[8]http://www.efsa.europa.eu/en/stakeholders_efsa/participating.html. Accessed 10 April 2007.

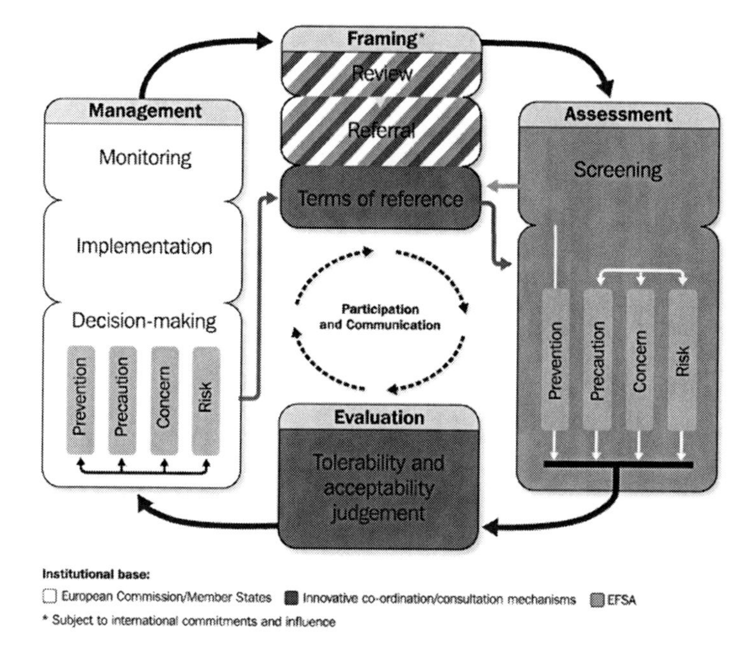

Fig. 2.8 A detailed representation of the General Framework, including the institutional allocation of tasks

feedback from later stages of the governance cycle to earlier stages, so that improvements can be made and problems averted. Specific examples of where this may be appropriate include:

- The possibility of reframing assessment – through the formulation of additional or altered terms of reference – following evaluation;
- The identification of gaps in knowledge about threats at the stages of evaluation or management, which will require further assessment to be carried out. In these cases terms of reference will need to be drawn up afresh through consultation and discussion within the Interface Committee.
- The identification of gaps in knowledge about management measures, which will necessitate targeted assessment by EFSA of the possible implications of these measures. Again, this will require the formulation of new terms of reference by the Interface Committee, with the opportunity of input from the Internet Forum.

Figure 2.8 illustrates the entire General Framework for food safety governance that has been presented above, including the various components of framing, assessment, evaluation and management, the cross-cutting activities of food safety communication and public involvement, the full set of possible interactions and feedback between all of these stages and the institutional bases to which the various tasks are allocated. The following chapters will discuss each of these stages, the cross-cutting activities, and the institutional implications in more detail.

Chapter 3
The Process of Framing

A. Ely, A. Stirling, F. Wendler, and E. Vos

3.1 Introduction

The previous chapter discussed various studies (Millstone, van Zwanenberg, Marris, Levidow & Torgersen 2004) which have highlighted the importance of risk assessment policy in influencing decisions around food and environmental safety. Risk assessment policy is the term used by the Codex Alimentarius Commission to describe

> documented guidelines on the choice of options and associated judgements for their application at appropriate decision points in the risk assessment such that the scientific integrity of the process is maintained (Codex Alimentarius Commission 2005: 46).

Codex views this as an activity that guides the scope and purpose of the risk assessment, for example by setting out the remit, who should participate, the questions that need addressing, how uncertainties should be dealt with, the factors that the assessors need to consider, the output form, and possible alternative outputs. From the point of view of Codex, risk assessment policy is a task to be carried out by risk managers. However, as set out in previous chapters, empirical insights into current practice of EU food safety governance have shown that risk assessment policy is a task already *de facto* shared between risk managers and risk assessors, with various initiatives of EFSA to take the lead to develop common approaches towards risk assessment, and to make risk assessment more harmonised and transparent. Therefore, it appears that EFSA has started to play a role in developing its own risk assessment policy, which is driven less by requests of risk managers than by its own priorities and insights (Vos & Wendler 2006b: 86). Against the background of these insights, the General Framework recommends that risk assessment policy should be understood as a task to be undertaken *jointly* by assessors and managers, in a fashion that is transparent to and takes account of inputs from a wide range of stakeholders. Risk assessment policy, according to the General Framework proposed here, therefore falls within the *process of framing*, which is carried out as a cooperative exercise within the interface between assessment and management, for which a specific institutional set-up is proposed by the General Framework (the details of which will be discussed in Chap. 6). The General Framework also recognises that this process of setting EU risk assessment policy involves, either

M. Dreyer and O. Renn (eds.), *Food Safety Governance*,
DOI: 10.1007/978-3-540-69309-3_4,

directly or indirectly, supranational organisations like Codex, as well as a variety of actors at national and EU levels.

Generally, the food safety governance activities represented by the framework in Fig. 2.8 are subject to various institutional and legal arrangements concerned with the assignment of responsibilities and the articulation of rights and obligations. The specific relevance of framing within this structure, as illustrated by the cyclical nature of the framework, lies in the fact that these processes are open to *design*, iterative *development* in the face of new learning, and to feedback between various stages in the process in response to regulatory *oversight* activities. The design and development of the process itself are guided by European Directives, Decisions, Regulations, and other European legal instruments and principles – which themselves can all become subject to change – and are moreover shaped by non-binding frames such as conventions, prominent perspectives and orientations, as well as by international influences. By explicitly including this as an element in the proposed framework within framing, it is acknowledged that the application of the precautionary principle as a general governance principle takes place at a number of organisational, legal and discursive levels, including institutional structure, process implementation and the exercise of administrative discretion.

Scholars in the fields of sociology and, in particular, science and technology studies (STS) have adopted the analytic term 'frame' or 'framing' to describe the ways in which individuals' or social groups' world views, or the conditions under which they operate, can influence the production and/or interpretation of data or knowledge (van Zwanenberg & Millstone 2005: 29; Jasanoff 2005: 23).[1] More recently scholars have applied this concept to empirical studies of science in policy making.[2] Within the process of framing described here, we identify a number of stages, which are described briefly below and illustrated in Fig. 3.1:

- *Review* – the ongoing process of adapting and improving the arrangements for food safety governance within the EU to respond to the global contexts in which they are situated. These contexts are made up not only of developments in scientific understanding (based in part on monitoring the effectiveness and consequences of existing management measures and on emerging upstream/basic research findings) but also of shifting socio-political, legal and institutional contexts at national, EU and supranational levels. Review does not apply to specific cases as much as to the regulatory structures within which these cases are dealt with.
- *Referral* – the process of referring a specific case (be it a new food product, production method, industrial process, or commercial practice) to EFSA for screening and later for assessment. According to the rules laid down in the

[1] The concept of a 'framing assumption' was first used by the sociologist Erving Goffman in Goffman (1974).

[2] See van Zwanenberg and Millstone (2005); Jasanoff (2005); Levidow, Carr, Wield, and von Schomberg (1997) and Wynne (1995).

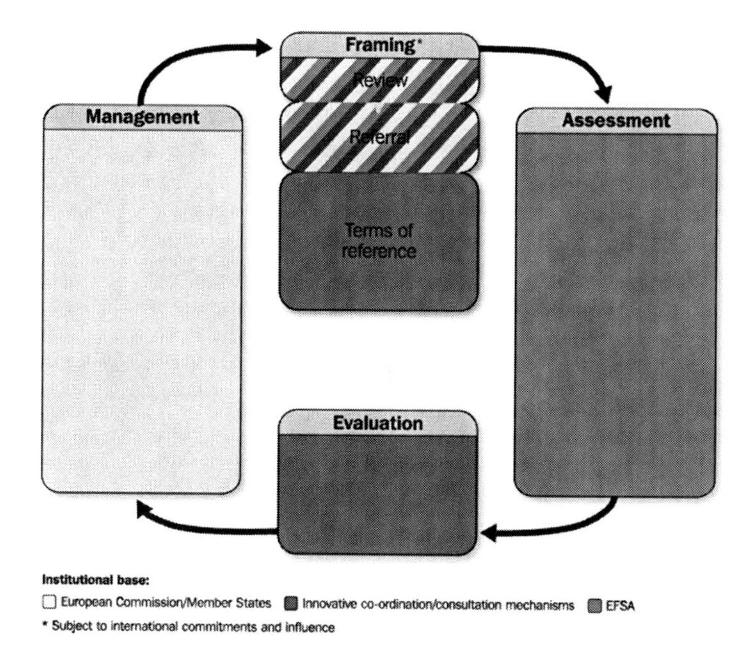

Fig. 3.1 The general framework, with an emphasis on the stages and institutional settings of *framing*

General Food Law, this may be carried out by the European Commission, by individual Member States, or, in the case of self-tasking, by EFSA itself.

- *Terms of Reference* – the process of setting detailed terms of reference, including information on the most appropriate assessment approaches for a specific case, upon which EFSA should act and issue a scientific opinion.

Each of these will now be dealt with in turn in order to outline the associated procedural arrangements and the salient aspects of their design.

3.2 Review

The term 'framing' will henceforth refer both to collectively binding rules and non-binding conventions and prominent perspectives. Within this process, we first identify and define the stage of *review*. Review describes the constant vigilance of regulators to new scientific evidence, technological developments, changing socio-political conditions or altered international regulatory frameworks and, subsequently, aims to produce timely responses to these dynamics. Thus, one aspect of review involves the legislative actors of the EU concerned with the formulation of binding rules in the form of European Directives, Decisions and Regulations which form the basis for the design of arrangements for handling specific products or processes (e.g. in terms of setting the rules for comitology procedures, defining the division of roles between Commission/EFSA, and setting out the respective functions and responsibilities of stakeholders and Member States). As such, the actors involved in the process of review are primarily the European Parliament, the Council, and the Commission (as the main actors responsible for legislative procedures), but also EFSA as the main responsible actor in the field of risk assessment. All of these actors are furthermore required to consider the input from a wide variety of European stakeholders. At the same time, international actors also have an input in this process of review.[3] The Codex Alimentarius Commission (CAC) is of major significance because of their increasing importance under WTO law. In this way, the sharing and transfer of emerging new scientific evidence and technological developments at international level are intrinsic to the conduct of review.[4] With regard to institutional arrangements, it is therefore clear that the factors affecting review cannot be identified with a single existing procedure or set of institutions, but refer to a range of processes and institutions that are relevant for setting the framework conditions of food safety governance. Above all, this is the adoption of framework legislation both of a general scope and individual legislative acts setting out procedural requirements for specific policy areas. Other relevant procedural arrangements that fit within the context of review are created not through full legislative procedures, but by single executive acts.[5]

[3] The main existing point of reference for this influence is the requirement set out in the GFL that international standards shall be taken into consideration in the development or adaptation of food law (Art. 5(3)). See in more detail Chap. 6.

[4] Review therefore includes an international aspect both by 'downloading' provisions and requirements established in international standards and by 'uploading' new developments and insights into the discussions and decision-making procedures at international level, and to ensure the compatibility between European and international developments. Apart from questions of compatibility of European food law with international standards, the GFL also establishes the obligation for the Community and Member States to contribute to the development of international technical standards for food, feed, sanitary and phytosanitary standards, and to promote the co-ordination of work on food and feed standards undertaken by international organisations such as the Codex Alimentarius Commission (Art. 13 GFL). See on interplay between Codex, EU and WTO law, Masson-Matthee (2007).

[5] Legal acts specifying the details of requirements and obligations set out in the GFL are often adopted through Commission Regulations. Examples for such decisions include Commission Regulation 2230/2004 EC on the networking of organisations operating within the field of mission of EFSA, or Commission Regulation 1304/2003 specifying the procedure for the handling of requests for scientific opinions by EFSA. Furthermore, many decisions concerning the involvement of stakeholder organisations and the realisation of principles of good governance are adopted through executive acts without the participation of the European Parliament, mainly on the basis of the requirements about the consultation of interested parties in the General Food Law (Articles 9 and 42). E.g. the creation of consultative bodies like the Advisory Group on the Food Chain was established through a Commission Decision, and the creation of EFSA's Stakeholder Consultative Platform, or the adoption of the Code of Good Administrative Behaviour of EFSA was made through a decision of the Authority's Management Board.

There are also examples of review which do not necessarily have a legal character, but are adopted in the form of declarations, guidance documents, or communications.[6]

Another aspect of review encompasses certain elements of risk assessment policy in that it influences or frames[7] the forms of knowledge that are gathered in the assessment process. Within the General Framework, assessment is further framed by a process (covered in detail in the next chapter) termed *screening*. Screening identifies the salient qualities of the products and defines processes around which knowledge needs to be gathered in order to ensure application of the most appropriate approach(es) to assessment. This not only ensures resources allocated to assessment being proportionate to the threats in question, but also helps to reduce potential negative impacts resulting from uncertainty, ambiguity, and ignorance (as defined in Sect. 1.1) by ensuring that the necessary levels of attention and appropriate methods are employed. As will be discussed further in Chap. 4, screening proceeds on the basis of set criteria, and the associated outcome will determine the approach to assessment taken. EFSA already has procedures and arrangements akin to this (although these may not be codified as such) that prioritise threats and allocate responsibility for their assessment to different Scientific Panels serving the authority.

In the General Framework presented here, the process of review includes those activities that govern the selection and characterisation of the threat criteria employed in screening. The General Framework allocates the responsibility for this part of review mainly to EFSA, on the basis of its tasks to promote and coordinate the development of uniform risk assessment methodologies, as well as to collect, analyse and summarise scientific and technical data in the field within its mission, and to take action to identify and characterise emerging risks (cp. GFL Articles 23 and 34). However, while EFSA is allocated primary responsibility for setting the criteria, it is suggested that the details of the criteria applied at the stage of screening should be included in the discussions taking place within the Interface Committee, thus allowing for inputs from risk managers and stakeholders.

Furthermore, review specifies the *relative priorities* attached to different threats and ensures that a *justifiable* and *proportional* balance is being struck in the allocation of resources to different aspects of screening, assessment, evaluation and management. In current practice, this set of tasks is undertaken by a variety of

[6]For example, efforts undertaken by EFSA (partly through self-tasking) to achieve a harmonisation of approaches towards risk assessment (in accordance with its tasks as defined in Art. 23 (b) in the General Food Law), which are more procedural in character and communicated through guidance documents and communications of the Scientific Committee.

[7]In the sense it was given by Jasanoff (2005), and van Zwanenberg and Millstone (2005).

actors within EFSA, the Commission, the Standing Committee on the Food Chain and Animal Health (SCFCAH), and the Parliament.[8]

The governance principles of participation, openness and accountability (CEC 2001a), and further commitments to good governance outlined in the General Food Law require transparent communication and the involvement of relevant stakeholders in each stage in the food safety governance process. The objective of fulfilling these principles during review could be served by making communications at each stage available to the public through a web-based forum (described in more detail in Chap. 6), managed by the Commission and providing a space for transparent input from risk managers, assessors, stakeholders, and citizens.

In general, it must be pointed out that review will necessarily involve a range of complex processes and a wide variety of institutions. It addresses any unforeseen difficulties that may arise and ensures that the overall framework is robust to changes in any circumstance. It also ensures that the process as a whole allows effective *social learning* to take place at every level, from the individual criteria to the architecture of the process as a whole. This allows for greater efficacy and efficiency, and, in particular, for the screening process to benefit from cumulative experience gained in assessment itself. The process should remain sensitive to wider evaluative and contextual issues and be open, from the outset, to engagement with the views and experience of different public constituencies and all interested and affected parties. In this context, consultations with interested parties during the preparation, evaluation and revision of food law (as required by Art. 9 GFL) may also constitute a part of review. Some of the discussions taking place at the level of the EFSA Stakeholder Consultative Platform already point in this direction.[9]

[8]Whereas the annual work programme of EFSA, essential for the prioritisation of threats, is adopted by the Management Board on a proposal from the Executive Director, the Management Board is required to make sure that both the annual work programme and the revisable multi-annual work programmes of EFSA are consistent with the Community's legislative and policy priorities in the area of food safety. Moreover, in drawing up the proposal for the annual work programme of EFSA, the Executive Director is required to consult with the Commission (GFL, Articles 25 (8) and 26 (2)). The prioritisation of threats is therefore influenced by both EFSA and the Commission. This has been confirmed by the information collected through interviews held with Commission officials, which revealed that the Commission increasingly consults with EFSA on the prioritisation of threats, instead of consulting with the Member States within the framework of the Standing Committee on the Food Chain and Animal Health. With regard to the allocation of resources to different aspects of food safety governance, however, the European Parliament (in co-operation with the Commission and Council) has significant influence through its control over the general budget of the European Union, which the budget for EFSA depends upon (cp. Art. 43 GFL).

[9]Examples include discussions about the general procedures and requirements for the provision of scientific advice (meeting of 9 March 2006), debates about risk communication strategies, transparency in risk assessment, and the identification and characterisation of emerging risks (meeting of 21 July 2006), or discussions about the working method of the Stakeholder Platform, the organisation of the interface with Member States and stakeholders, or EFSA's future work and priorities (meeting of 6 December 2006).

To sum up, in comparison to the other aspects of framing described below, review refers largely to what may be called the *meta-level* of food safety governance involving institutions outside the primary focus of this book (such as the WTO). As such it is – to limit the scope of the current exercise – largely excluded from the considerations of procedural and institutional challenges and possibilities for innovation. It still deserves, however, to be addressed in future research on how the innovations proposed here may be implemented.

3.3 Referral

The second stage in framing, *referral*, involves the forwarding of a particular case to EFSA for assessment, usually with reference to a particular law under which the associated threat(s) should be assessed and managed. The General Framework proposes that details of the referral, including the legal jurisdiction under which the case is referred to EFSA, should be presented transparently on the Internet Forum (for specifications, see Chap. 6). Through this exercise, space should be made available for comment, which can be taken into account during the stage of screening (carried out by EFSA) and the final stage of framing, the setting of terms of reference. In current practice, the task of referring cases to EFSA is already structured by a variety of legal requirements and provisions. The main requirements for the referral of cases to EFSA are set out in Art. 29 of the GFL, which entitles the Commission, the Member States, and the European Parliament to request scientific opinions, and EFSA to issue opinions on its own initiative. The exact procedures to be applied in the handling of such requests are set out in Commission Regulation 1304/2003, which *inter alia*, recommends that such requests be made in an objective, transparent and functional manner.[10] The proposal of the General Framework to make the referral of cases to EFSA more transparent through the publication of draft terms of reference in the Internet Forum, builds on this objective. Regulation 1304/2003 stipulates that in all requests for scientific opinions, it is essential for the applicant to remain responsible for the substance of the question posed and to agree to any amended request before it is forwarded to the scientific committee (CEC 2003, Recital 6).

It is therefore clear that if screening adds additional insights to a case referred to EFSA, and the exact terms of reference are only agreed on after the results of screening have been discussed within the Interface Committee, the applicant for a request would by law have to participate in the drafting of the terms of reference and agree to the final version passed on to EFSA. In addition, many cases are referred to EFSA on the basis of case-specific legislation, such as in the authorisation procedures for

---

[10] Commission Regulation 1304/2003 EC of 11 July 2003 on the procedure applied by the European Food Safety Authority to requests for scientific opinions referred to it, OJ L 185/6 (CEC 2003, Recital 5).

genetically modified food and feed (specified in Regulation 1829/2003 EC) or the authorisation procedure for food contact materials (specified in Regulation 1935/2003 EC). In both instances, cases are referred to EFSA within a specifically prescribed authorisation procedure and accompanied by full technical and scientific dossiers, as prescribed by the relevant legislation and guidance documents. Therefore, these cases may differ from cases which are referred to EFSA asking for a scientific opinion about an emerging threat or a question of a more general nature, such as the request by the European Parliament for a scientific opinion on wild and farmed fish (EFSA 2005). In practice, a large part of the cases in question are referred to EFSA by one of the Member States on the basis of these authorisation procedures, i.e. following the request of a private applicant – mostly enterprises wishing to place their products on the markets – in one of the Member States, instead of questions from a national food safety authority.

Therefore, the conditions (e.g. legal context) under which referral takes place can also frame the way in which the assessment will be carried out. EFSA then proceeds with the screening of the threat, informing the most appropriate form of assessment, which is then specified further by the setting of terms of reference.

3.4 Setting the Terms of Reference

Once screening has identified the most salient characteristics of the threat at hand, the detailed *terms of reference* upon which the assessment should be based need to be defined. In the current practice, this is usually done by the European Commission. As described earlier (cp. Sect. 1.2.2), our analysis has indicated that there is a need for enhanced co-ordination between managers and assessors in this activity. The governance framework as advocated in this book, envisions the terms of reference to be set in a transparent way jointly by these two actors in cooperation with key stakeholders (through the Interface Committee). Furthermore, the proposed framework would see the draft terms of reference displayed in the Internet Forum in order to provide affected and interested actors with the possibility to give input (for details concerning the tasks and structures of the Interface Committee and the Internet Forum, see Chap. 6). Under current structures, EFSA is legally required to establish a register of requested opinions which is accessible to the public, allowing the progress of requests for opinions to be followed from the date on which they are received (CEC 2003, Art. 2). Although this register of scientific opinions is accessible on the EFSA website,[11] the terms of reference of ongoing risk assessments cannot be retrieved from this register and are only made public *ex post* as part of established opinions of EFSA available through the register. By also making public the draft terms of reference 'in real time', and allowing stakeholders and interested parties to comment on them, would allow the Commission to make use of this input, and to respond to it.

[11] See: http://www.efsa.europa.eu/register/qr_panels_en.html. Accessed 10 May 2007.

The institution taking responsibility for the final terms of reference should justify the text chosen, based on a summary of the various points made on the Internet Forum and including any constraints or requirements emanating from the stages of review or referral, discussed above. This summary should be published on the Internet Forum as an accompanying document to the final terms of reference. Following the issuing of the final terms of reference by the Commission or the Interface Committee (see Chap. 6), which have been formulated through interface communication between the different parties, EFSA continues with its established role of assessment. This process is the subject of the following chapter.

Chapter 4
The Process of Assessment

A. Ely and A. Stirling

4.1 Introduction

This chapter is dedicated to those activities carried out solely by assessors, largely EFSA, focusing on the work of EFSA under the proposed General Framework. As has already been mentioned, the first activity, *screening*, involves the identification of the most appropriate assessment approach for the threat in question. Detailed criteria for screening threats are developed during the process of review, as mentioned briefly in the previous chapter; the actual use of these criteria will be treated in more detail in Sect. 4.2, below. The various aspects of the actual *assessment* process, how they relate to the legal and institutional requirements of good governance outlined in Sect. 1.2.2, and how they can help to overcome the challenges outlined in Sect. 1.3, will then be addressed in Sect. 4.3.

Prior to addressing the function of screening, it is necessary to introduce the different approaches to assessment that are understood within the proposed framework. The distinguishing characteristics of exactly what constitutes conventional risk assessment tend to vary slightly between different intergovernmental and European Commission definitions. The particular stages of conventional risk assessment recognised in European regulation of food safety comprise: hazard identification, hazard characterisation, exposure assessment and risk characterisation (GFL, Art. 3). In common with similar understandings throughout the field of safety regulation worldwide, this embodies the central understanding that risk assessment involves the use of probabilistic techniques to address incertitude over the likelihood of different possible outcomes.

Despite its prominence – in the field of food safety as elsewhere – conventional risk assessment does not present the only methodological approach to assessing different products, processes or policy options (Yapp et al. 2005). Indeed, depending on the context and conditions, a number of alternative or additional methods can offer *more comprehensive* approaches to assessment than is achievable using conventional risk assessment. For instance, procedures such as horizon scanning, sensitivity analysis, interactive modelling, and scenario workshops provide more comprehensive means to represent and examine the range of possible outcomes without aggregating them together. Likewise, analytic-deliberative processes of

M. Dreyer and O. Renn (eds.), *Food Safety Governance*,
DOI: 10.1007/978-3-540-69309-3_5,

decision analysis, multi-criteria mapping, stakeholder engagement and citizen participation can identify a more comprehensive range of questions, options, assumptions, and values and allow fuller exploration of their effects on the outcomes of assessment, than are usually addressed in conventional risk assessment.

Together with more quantitative approaches focussed on risk, these techniques offer a rich and powerful array of possible approaches to assessment. Each individual approach – and a host of variants, composites and hybrids – displays contrasting characteristics in relation to different principles of good governance. There can be significant tensions and trade-offs between qualities such as timeliness and proportionality, on the one hand, and accessibility and effectiveness, on the other, or between the imperatives for participation and accountability and those for coherence and consistency. Different approaches are favoured under divergent institutional, disciplinary and socio-political perspectives. It is clear that no one assessment approach offers a panacea for all possible empirical contexts or governance conditions. But it remains unclear how best to go about reconciling the tensions, trade-offs and perspectives in order to identify the most appropriate approach to take, under any given context or condition.

The use of the term *threat* in this framework is important for purposes of consistency and coherence. It was explained in Sect. 1.1.2 that the scientific definition of the term *risk* implies conditions under which both probabilities (exposures, frequencies) as well as magnitudes may lend themselves to quantification. As such, it is conventionally distinguished from a "hazard", for which only magnitudes (in terms of potential for damage, without considering exposure or probability) may be characterised with confidence. The term threat, which is also used in influential governance instruments and documents,[1] is chosen because it covers *both* risk and hazard and admits interpretation either in terms of probabilistic risk or intrinsic hazard properties, depending on the context. Screening is therefore focused on threats including hazards and/or risks depending on knowledge and context. For many regulatory purposes such as determining maximum daily intakes, empirical data on exposure is not important so that hazard information is sufficient for the assessment and management process to follow.

In the field of food safety, examples of intrinsic hazard properties may relate to endpoint effects (such as cancer, genetic disorders or allergies) or to exposure potentials (like bioaccumulation, persistence, and ubiquity). Either way, the screening of threats involves attention to the basic elements of precaution (seriousness and lack of scientific certainty) as well as additional considerations concerning the socio-political ambiguity of the threats in question. This requires sets of operational criteria for triggering the different assessment approaches that are discussed in more detail in Sect. 4.2.

[1] For example, Principle 15 of the Rio Declaration on Environment and Development.

4.2 Screening

4.2.1 Screening Criteria

What is here termed the screening of threats corresponds approximately to established notions of hazard identification, basic characterisation and "preliminary risk assessment", as featuring, for instance, in discussions under the auspices of the WTO and elsewhere. This requires that a systematic and transparent approach, which can be either quantitative or qualitative, be adopted to the achieving of two main aims: First, to guide the allocation of different broad types of threat to the most appropriate, efficient and proportionate form(s) of assessment; second, to inform the prioritisation of attention and resources in assessment to different instances of threat within these broad types. The two tasks are closely interlinked, since information gained during screening for the first aim is also likely to be useful in addressing the second.

In order to meet the challenges identified in Chap. 1, a number of further specific attributes of a threat must be clearly addressed in the screening process. In particular, the following elements must all be systematically scrutinised in this process: the level of *seriousness* of a threat; the extent to which it is subject to scientific *uncertainty* and the levels of socio-political *ambiguity* with which it is associated. Each implies the necessity of different kinds of information in the subsequent assessment process. In the General Framework, efficient and effective allocation to these different assessment processes is achieved by means of a series of explicit criteria, against which each threat in question is examined. The adoption of particular criteria will depend, in part, on the legal and regulatory context (included within the review stage of framing) and will be subject to normal provisions for design, development, and oversight. While the criteria outlined in this book are broad enough to be applied to most food safety threats, more specific and detailed criteria could be drawn up by the Interface Committee that would relate to particular types of food products or processes, or be designed so as to be applicable under certain food regulations. The involvement of managers, assessors and stakeholders on the Interface Committee (as well as use of the Internet Forum) will provide for co-ordination, as well as openness and transparency in the setting of these detailed criteria. It is suggested that the Interface Committee regularly reviews these criteria which may lead to their reformulation.

Under each criterion, some threshold level or characteristic is established, which identifies this threat as registering under that criterion. This is then taken as a basis for assigning this threat to a particular form of attention in subsequent assessment. In this way, the application of successive criteria serves clearly and consistently to allocate particular types of threat to particular forms of regulatory treatment. Additional information gained in this screening process will be very useful in the prioritisation of attention to the different types of threat *within* the different assessment procedures.

Of course, the application of the criteria that inform the screening process is not purely mechanical. There are typically close inter-relationships between criteria, requiring that they be applied as part of an integrated, reflective, deliberative process, accountable to the appropriate institutions of design oversight. A general working sequence is suggested from seriousness to precaution with ambiguity being somewhat separate and considered in parallel to precaution. In other words, in the interests of effectiveness and proportionality, the question as to whether a given threat is "certainly and unambiguously serious" is clearly prior to the other considerations. Only in the event that the response to this question is "no", does attention turn in sequence to the various reasons why this might be the case.

A negative response to this initial question of seriousness may variously be because the threat in question is scientifically uncertain, socio-politically ambiguous, or is certainly and unambiguously *not* in excess of the chosen criteria of seriousness. Of course, where a particular threat displays multiple attributes, for example conforming to screening criteria for both ambiguity and uncertainty, then these different aspects may be treated *in parallel* by different forms of assessment.

4.2.2 Criteria of "Seriousness"

The first step in the screening process is therefore to identify whether the threats in question are "certainly and unambiguously serious". Subject to further findings in the parallel review of existing institutional practice our team has developed a number of specific exposure-based hazard criteria for general application to food safety threats. These include *carcinogenicity*, *mutagenicity* and *reprotoxicity* in food components or residues (as already embodied in existing regulatory initiatives in this field, such as the 2001 European Commission's Chemicals White Paper, CEC 2001b). Beyond this, attention may extend to further health threat criteria such as *endocrine disruption*, *neurotoxicity*, *asthmagenicity* or *sensitising potential*. In other contexts, threat criteria might be formulated in terms of other types of food safety hazard, such as the presence of certain particularly virulent *pathogens* or the inclusion of those *antibiotic resistance* marker genes that were opposed in genetically modified organisms by the EFSA Scientific Panel on Genetically Modified Organisms in 2004 (EFSA 2004b). Alternatively, in areas where there exist robust applicable data, threat criteria may be formulated in terms of risk-based thresholds, such as *concentrations* for certain less hazardous pathogens or toxicants.

As has been noted, these criteria are all subject to discussion as part of the review stage of the framing exercise. Prevention is then chosen when examination of the threat based on these criteria leads to the conclusion that it violates an existing legal requirement, exceeds a threshold of previously established standards or norms (based on a legal or institutional requirement to act) or is highly likely to exceed such a threshold. In addition, if a new threat is found where analogies to existing intolerable threats can be drawn, the presumption of prevention is justified.

Such a judgement may be obvious in many cases and uncontested; in other cases there may be dissenting views or differences in opinions. If that is the case, one of the other three assessment approaches has to be taken. The first criterion combines two qualifiers: the threat has to be serious and the judgement has to be univocal. When both conditions apply, then *preventive measures* are triggered.

4.2.3 Criteria of "Scientific Uncertainty"

In considering whether a threat is certainly serious under criteria such as those identified above, an accompanying step in the screening process is to identify specific criteria for what constitutes "scientific uncertainty". A crucial issue here concerns the applicability of probabilistic risk assessment techniques. As outlined in Sect. 1.1.2 above, difficulties in this respect may lie not only in addressing *uncertainty* (where by definition, we cannot confidently derive probabilities for at least some sub-set of outcomes), but also *ignorance* (where some outcomes themselves may be entirely unanticipated).

Our team has developed a series of candidate criteria for identifying all these forms of scientific uncertainty which are not fully characterisable by probabilistic techniques. The first two address different aspects of ignorance, insofar as this is possible, by focussing on sensitivities to the prospect of surprise. The remaining criteria address different aspects of uncertainty. Taken in logical sequence, the criteria are as follows:

(a) Are there scientifically founded questions concerning the status of the theoretical foundations of the disciplines bearing on the characterisation of the threat?
(b) Are there features of the food or food component in question which are substantively novel, in the sense that they involve characteristics or properties that are in some sense unprecedented?
(c) Are there scientifically founded questions concerning the completeness or sufficiency of the particular scientific models bearing on the characterisation of the threat?
(d) Are there scientifically founded questions concerning the applicability to the context in question of the particular scientific models used to characterise the threat?
(e) Are there scientifically founded questions concerning the applicability to the context in question of the data-sets bearing on the characterisation of the threat?
(f) Are there scientifically founded questions concerning the quality of the data-sets bearing on the characterisation of the threat of a kind that is not susceptible to probabilistic treatment?
(g) Do there exist any indirect, interactive or synergistic causal mechanisms of a kind that may not fully and confidently be characterised by probabilistic techniques?

Where a consensus does not emerge between the EFSA personnel responsible for screening as to the presence or absence of uncertainty (as defined by the above criteria), it is assumed that the high level of protection would lead to an assumption of uncertainty, as if one of the above criteria was triggered. Where they are held to be acceptable in principle, such criteria can be elaborated further by reference to an extensive existing literature. Where any one of them exceeds predefined quality criteria (pertaining to deficits in theory and modelling) or limits of foreseeable variability (pertaining to data analysis and interpretation, for example by using Monte Carlo-simulation techniques), then the threat in question is assigned to *precautionary assessment*.

4.2.4 Criteria of "Socio-Political Ambiguity"

In addition to the initial screening question over scientific uncertainty, the other reason why threats may be identified *not* to be definitely serious is where they are socio-politically ambiguous. This focuses on the degree to which a given threat may be subject to strongly divergent cultural attitudes, political perspectives, or economic interests. There are four types of criteria that can be used to identify these kinds of ambiguity.

(a) At the level of individual constituencies: is there a perceived threat of harm on a catastrophic scale (*individual criterion*)?
(b) Where there is disagreement between regulatory agencies and/or Member States: are there aspects of these institutional conflicts ostensibly unrelated to scientific uncertainty (*institutional criterion*)?
(c) With regard to the news media: are there signs that the threat in question is subject to a pronounced degree of amplification (*amplification criterion*)?
(d) At the level of society as a whole: are there signs of adverse effects in terms of social justice in the distribution of threat, or in terms of manifest political mobilisation on the part of particular public constituencies (*social criterion*)?

Where any one of these criteria applies, then the threat in question is assigned to a process of *concern assessment*.

4.2.5 Threats Not Addressed by Above Screening Criteria

Where a threat is found not to be serious, uncertain, or ambiguous under any of the screening criteria described so far, then it will by definition trigger criteria for the applicability of conventional risk assessment (meaning that probabilistic techniques are applicable). Such threats are best addressed by drawing on a variety of risk assessment techniques, depending on the nature of the problem at hand.

Under circumstances where an extensive epidemiological record of safe use exists, then *standard risk assessment* may be appropriate. This usually involves the simple combination of hazards (as characterised through dose–response relationships, for example) and exposures (as evident from established data sets). At other times, a more *extended risk assessment* may be required. In these cases, conventional probabilistic techniques may still be applicable, but need to be applied in a more wide-ranging and elaborate fashion than is normally the case.[2]

The kinds of threats necessitating extended risk assessment are complex (if the threat is subject to complex cumulative or additive causal mechanisms) or large in scale (if a number of people exposed exceed a certain threshold). In addition extended risk assessment may be required if the maximum possible harm exceeds a certain threshold magnitude or if the time lapse between the policy decision in question and the manifestation of the resulting impacts exceeds a certain threshold time period (for example in the case of intergenerational effects). If the response to any of these questions is uncertain, then this should already have been picked up in applying the uncertainty criteria specified above. However, the finding of particular reasons for uncertainty at this stage might prompt re-application or re-interpretation of the earlier uncertainty criteria in light of the new evidence.

4.3 Assessment

The purpose of assessment is to gather the information necessary to inform and substantiate a particular governance outcome.

The type, scope and quality of information relevant to this decision making will vary from context to context and from threat to threat. Depending on the context and magnitude of the threats in question, it may be necessary to include assessment of socio-economic as well as health factors. In the interests both of efficiency and effectiveness, it is desirable for the terms of reference (informed by screening, above) to be as specific as possible about the most appropriate form to be taken by the assessment process in any given context.

---

[2] The third subproject of the SAFE FOODS project has adopted probabilistic techniques to model the health impacts on European populations to pesticide, mycotoxin and natural toxin exposures. Where probabilistic risk assessment is applied, it should not be used inappropriately as an aggregative tool exclusively to justify or enforce ostensibly definitive monolithic claims to safety or to the unitary sufficiency of intervention measures. *Sensitivity analysis* (both analysing the effect of data and model uncertainty on the assessment) is an essential part of such quantitative techniques and is recognised as such by other subprojects in SAFE FOODS. While subproject 3 has reported adequate data in relation to pesticides, data on mycotoxins and natural toxins have been poor both in availability and quality (Subproject 3 report-back session, SAFE FOODS Consortium Meeting, Pretoria, South Africa, 25 May 2006). Especially under such circumstances, where the scarcity of data means that assessment must be assumption- (rather than data-) driven, uncertainty criteria may in addition be triggered (necessitating a precautionary approach to assessment).

Instead of a single undifferentiated notion of "risk assessment", then, the present framework distinguishes *four different approaches to assessment* (corresponding with the four potential outcomes of screening). In the terms alluded to in the existing General Food Law, as reviewed in Chap. 1, the more elaborate forms of assessment detailed here each represent a different specific way in which assessment might be "more comprehensive" than standard risk assessment. The four different approaches to assessment, and their relationship with framing and with the screening process described above, are illustrated in Fig. 4.1 below.

4.3.1 Presumption of Prevention

Where threats are identified in the screening process certainly and unambiguously to be serious (illustrated by the question "serious?" in Fig. 4.1), then the presumption is that they are assigned directly to preventive measures. Here, assessment simply involves consideration of whether there exist any *mitigating factors* that justify *conditional relaxation* of restrictive regulatory instruments. Such mitigating factors may take the form of countervailing risks, overriding benefits or unavoidable constraints on control.

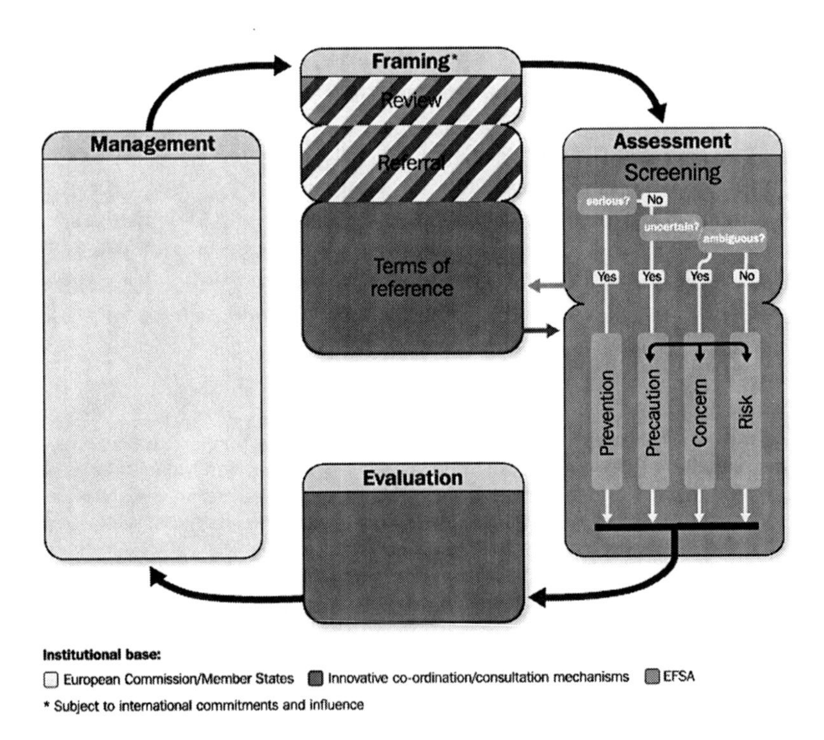

Fig. 4.1 The General Framework, with a focus on the stages of *screening* and *assessment*

In those rare cases where prevention is argued to be counter-balanced by such mitigating factors, then this effectively implies that the triggering of criteria of "certain and unambiguous seriousness" is, in this particular instance, correspondingly qualified. Depending on whether the qualification takes the form of uncertainty or ambiguity, the threats in question will be assigned for further attention either (respectively) to precautionary assessment or concern assessment. In either case, the presumption of prevention will be augmented by critical examination of such potential mitigating factors or grounds for conditional relaxation as part of a comprehensive and inclusive deliberative process, involving relevant interested and affected parties. Such rare instances should also be subject to particular attention as part of the overarching framing process.

Under a presumption of prevention, assessment of socio-economic factors is included alongside more direct issues of hazard and risk as a means to inform judgements over the nature of any "countervailing risks, overriding benefits or unavoidable constraints on control".

4.3.2 Key Features of Precautionary Assessment

Where the identification of a threat displays a lack of scientific knowledge about probability distributions and/or the magnitude of harm (illustrated by the question "uncertain?" in Fig. 4.1), then the presumption is that the product, process or practice in question will be subject to precautionary assessment. This does not automatically imply the implementation of preventive measures. A wide variety of regulatory measures may result. In essence, precautionary assessment involves more detailed and broader-based consideration of the factors bearing on the threat in question and a *comparative review* of a set of functional equivalents to the product/process/practice in question.

Here (recalling the discussion of different forms of incertitude in Sect. 1.1.2), a practical distinction can be made between *institutional ignorance* (located specifically at the point of decision making) and *societal ignorance* (a generic property of the state of knowledge extant in society as a whole). The former can be addressed by "broadening out" the assessment process in the ways detailed in the criteria below. This ensures that as much pertinent knowledge and experience as possible is brought to bear on decision making. Beyond this, a number of other provisions can directly address the more intractable latter forms of societal ignorance. A series of key characteristics can be identified:

(a) Extension of the scope of assessment to include a range of *indirect* forms of exposure, *additive*, *cumulative* and *synergistic* effects occurring throughout the food chain, addressing *mixtures*, *derivatives* and *reaction products* that may be present in final foodstuffs as well as considering institutional *trends* and *complianceissues*. These aspects are part of a precautionary assessment if the causal connections are not well understood and cannot be modelled with a high degree of confidence in an extended risk assessment.

(b) Address aspects of *institutional ignorance* by engaging a full range of technical *disciplines* and *stakeholders* right at the outset in assessment, in order to elicit the pertinent *prioritisation*, *conceptualization* and *interpretation* of the different questions that may be posed of the scientific data and the comprehensive exploration of the resulting *sensitivities*.
(c) The systematic examination of the potential adverse effects for public health associated with the products, processes or practices presenting the threats in question at the *earliest stages* in the innovation process.
(d) Subject to the terms of reference, the detailed and balanced comparison of contending merits and drawbacks of a series of strategic options which present *alternatives* – in the sense of functional equivalents – to the product, process or practice in question, including inaction and the status quo and better ways to provide the goods or services in question. This includes the eliciting of the knowledge and also the concerns and preferences of stakeholders regarding the different alternatives and their social and economic implications.
(e) A shift in the *burden of persuasion*, such that it is those wishing to implement the technology or product in question who must resource the acquisition of relevant data and sustain an argument as to the acceptable nature of the associated threat, subject to an appropriate level of proof.
(f) An explicit focus on the extent to which the technologies or products under scrutiny display properties of *flexibility*, *adaptability*, *reversibility* and *diversity* – all of which offer different ways of hedging against exposure to any residual societal ignorance that has not been addressed by the other elements in precautionary assessment.

These elements of precautionary assessment are best addressed by taking into account all relevant bodies of knowledge, including that available from different natural and social scientific disciplines, as well as experiential knowledge on the part of different organised interests and groups such as workers, consumers, or local residents. Where socio-economic, as well as scientific uncertainty exists – for example, when the potential outcomes for the livelihoods of various sections of society, or the impact on the broader economy cannot be predicted with confidence – similar techniques to those listed above may be applied to the assessment of socio-economic risks and benefits. This generally relates to a broadening out of the assessment process to a wider range of disciplines and stakeholders, a shift in the burden of persuasion to those who wish to implement the technology or product in question, and a balanced comparison of strategic options in order to gather information on the relative benefits and risks of various functional equivalents.

Precautionary assessment is based on knowledge (systematic and experiential), not on beliefs or value judgments. That is why participation in the resulting analytic-deliberative exercise should be limited to *knowledge acquisition*. Examples of processes for eliciting stakeholder knowledge might include hearings, focus groups, or surveys.

4.3.3 Key Features of Concern Assessment

Where a threat is identified not to be definitely serious under the chosen criteria, nor subject to scientific uncertainty, but where screening has identified socio-political ambiguity (illustrated by the question 'ambiguity?' in Fig. 4.1), then the choice of appropriate management measures will be subject to a process of concern assessment designed to clarify and help resolve this ambiguity. The available methods for concern assessment take a variety of forms:

(a) The commissioning of large scale quantitative surveys, focusing as appropriate on representative, weighted or particular relevant groups.
(b) The conduct of qualitative social scientific procedures such as focus groups, examining the perspectives of specific sensitive or exposed groups.
(c) The design of extensive expert Delphi procedures in which a diverse array of interdisciplinary specialisms are focused on resolving the relevant questions.
(d) The direct retaining of wider social science expertise to observe, engage with and explain processes of social mobilisation.
(e) The holding of formal hearings with relevant social interest groups or targeted at relevant public constituencies as a means to elicit their concerns (such as affected local communities).
(f) The convening of deliberative bodies such as trans-disciplinary commissions to elicit as wide a range of concerns, visions, and mental associations as possible.

The above methods may be applied to the assessment of ambiguous socio-economic impacts as well as those dealing directly with human health issues. Relevant examples might include instances in which certain outcomes deliver disproportionate benefits to certain sectors of society but impose risks on other groups who do not stand to gain. In any event, the choice of appropriate methods for the process of concern assessment will itself be a matter for careful deliberation on a case by case basis. This will necessarily be closely interlinked with the activity of review (involving design, development and oversight of the food safety governance structures within which these cases are attended to) and the setting of the terms of reference.

4.3.4 Conventional Risk Assessment

Where threats are identified in the screening process as neither characterised by unresolved uncertainty nor ambiguity, the presumption is that they are subject either to deterministic or (in the case of modelled uncertainties) probabilistic risk assessment procedures. In cases of standard risk assessment, assessment takes a *straightforward form*, based simply on *probabilities* and *magnitudes*, and is performed by panels of independent experts, assisted by staff from the regulatory bodies concerned.

There is no particular need for involvement by external actors. If this routine process identifies any residual uncertainties, ambiguities or complexities that may have been missed in screening, then the threats are referred to one of the more comprehensive assessment procedures, as appropriate. Of course, this assessment process, as are the others, is subject to general political oversight and accountability.

Extended risk assessment involves detailed consideration of all aspects of the threat in question, including systematic modelling of different exposure pathways, with their associated probabilities. This allows the determination of appropriate safety margins. The process is undertaken in a fully transparent and accountable fashion by interdisciplinary groups of specialists, with full independence from special interests and external to the regulatory bodies concerned. Particular attention is directed at the factors identified under the criteria discussed above: the complexity of the causal mechanisms, the number of people exposed, maximum extent of possible harm, and the time lapse between the commitment and manifestation of effects. If uncertainties remain beyond the level of acceptable confidence intervals, then the risk is referred to a precautionary approach. Where justified by the relevant expertise, conventional risk assessment may also involve scientific engagement by experts from stakeholder groups.

Under conventional risk assessment, the priority attached to consideration of socio-economic factors will depend on the context and magnitude of the threats in question. Where assessment reveals risks to be low in magnitude, then – as at present – it would not be efficient or proportionate to include detailed assessment of socio-economic factors. However, as the magnitudes of risks are recognised to increase, there will be a corresponding necessity to provide subsequent evaluation and management stages with information concerning the nature and scale of any socio-economic benefits or justifications for the toleration of what might otherwise be seen as relatively high levels of risk.

A scientific colloquium held by EFSA in 2006 suggested that a favoured basis for future practice under such conditions might incorporate the definition of a common scale of measurement (e.g., disability-adjusted life years or DALYs, quality-adjusted life years or QUALYs, or, even more simply, Euros) for comparing the risks and the benefits of particular risk management measures (EFSA 2006b). It remains for EFSA formally to adopt an approach for this purpose. The complexities involved in assigning unitary measures to outcomes which may be subject to divergent evaluations by differing stakeholder groups make this approach particularly vulnerable as a tool on which to base policy. Bearing in mind the weaknesses of such reductive quantitative approaches, the appropriateness of alternative analytic-deliberative processes should not be understated. Decision analysis, multi-criteria mapping, stakeholder engagement and citizen participation – which may be drawn upon alongside other social scientific elicitation techniques in the process of concern assessment – can help to open up assessment to some of the socio-economic dimensions of food safety decisions whilst avoiding the over-simplification of aggregative techniques.

4.4 Potential Opportunities for Interlinkages Between Different Forms of Assessment

Potential *interlinkages* exist between the approaches of precautionary assessment, concern assessment and conventional risk assessment. The opportunities for interlinkages between different forms of assessment will of course depend on the specific features of the case in point. One specific threat may have impacts that demand extended risk assessments (for example health risks) and other types of impacts that would suggest a precautionary or concern approach (for example looking into environmental impacts or ethical implications). The different approaches are not mutually exclusive but can be *combined* depending on the nature of the threat and the different types of impacts under review. The opportunity for interlinking different forms of assessment may be specified in the terms of reference, or alternatively may be initiated by the assessors themselves.

It is important to stress that the assessment process may also reveal errors resulting from the screening process. For example, a threat may have been routed to the extended risk assessment approach but, during the assessment, it may become obvious that a precautionary approach is more suitable. It is therefore essential that during the assessment process *checks* about the need for re-routing to another approach are incorporated in the assessment process.

4.5 Outputs of Assessment

Following the principle of transparency put forward in the other stages in the food safety governance cycle, the outputs of assessment and the supporting documentation should be made available on the Internet Forum, to allow comment and feedback – and, where necessary, challenge – by stakeholders and citizens. Where such deliberation uncovers issues that were not adequately addressed in assessment, these issues can be referred back to the EFSA for screening, after which new terms of reference can be formulated in order to address them adequately.

Following the process of assessment, in which knowledge in various forms is accumulated in order to inform decision making, the governance framework proposes the two processes of evaluation and management. It is here that the knowledge is assimilated, and stakeholders' values brought to bear on the outputs of the assessment process so that scientifically informed and democratically accountable decisions can be made. The next chapter addresses both the processes of evaluation and management.

Chapter 5
The Processes Evaluation and Management

O. Renn and M. Dreyer

5.1 Introduction

The main purpose of the *evaluation* stage is to judge the *tolerability* or *acceptability* of a given threat and, if deemed necessary, to initiate a management process. The chief purpose of the stage of *management*, closely related to the stage of evaluation, is to decide on *intervention measures* which will range in each case from strict prohibition (such as bans and phase outs) to unrestricted permission. In between, there lies a wide range of measures, including legal requirements (such as exposure standards, engineering regulations, and best practice), financial instruments (such as mandatory insurance, assurance bonds, or tradable licenses), private self-regulations (such as in-house quality control) and information and educational strategies (such as consumer information, labelling, and classroom curricula). Following a regulatory impact assessment of the possible measures, investigating their feasibility to and acceptability by stakeholders, one or more appropriate measures are selected and implemented, and enforcement details and options for review are determined. The various key features of evaluation and management are illustrated in Fig. 5.1 below.

There is no intrinsic correlation between each respective approach to assessment and particular evaluation and management procedures, or management measures adopted. However, depending on whether a given threat is characterized as definitely serious and cannot be justified by any mitigating factors, as a scientifically uncertain threat, or as a socio-politically ambiguous threat, certain procedures and measures are *especially suited* for handling the threat in evaluation and management.

M. Dreyer and O. Renn (eds.), *Food Safety Governance*,
DOI: 10.1007/978-3-540-69309-3_6, © Springer-Verlag Berlin Heidelberg 2009

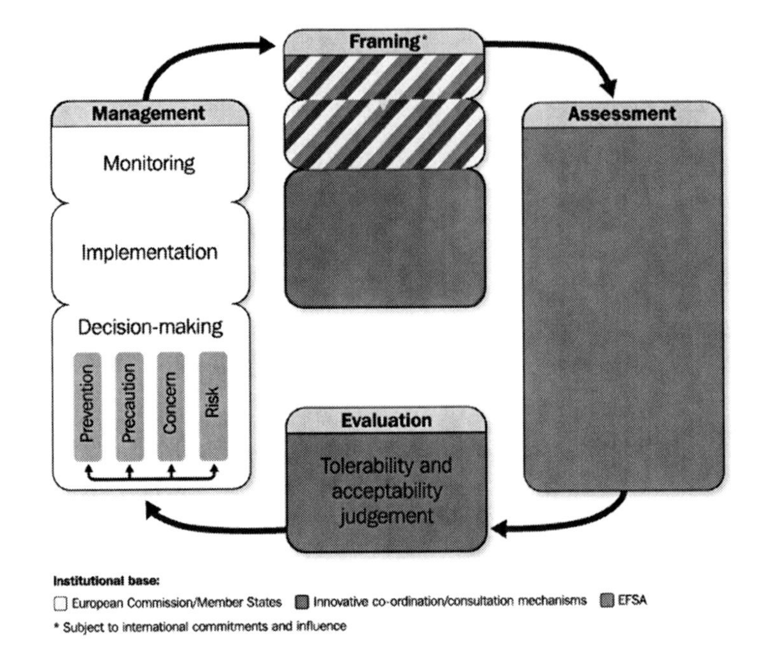

Fig. 5.1 The General Framework, with a focus on the stages of evaluation and management

5.2 Tolerability/Acceptability Judgement in Evaluation

The step of evaluation, which follows after the assessment stage, implies that the insights of the assessment exercise are summarised and deliberated in consideration of wider social and economic factors in order to inform a decision on the necessity of intervention measures and the selection of appropriate management measures.

While assessment deals with knowledge claims (around what are the causes and what are the effects), evaluation deals with *value claims* (around what is good, acceptable, and tolerable). Assessment is about collecting and summarising all relevant evidence necessary for making an informed choice on the threat's tolerability or acceptability; evaluation means applying societal values and norms to the judgement on tolerability and acceptability and, consequently, determining the need for management measures. The tolerability or acceptability judgement is informed, but not determined by the results of the assessment process. It will be based on balancing *pros* and *cons*, testing potential impacts on quality of life, discussing different strategic options for economy and society, and weighing the competing arguments and evidence claims in a balanced manner.

The outcome of evaluation might lead to further systematic scientific assessments, beyond that of health effects, being commissioned to outside institutions with the required special expertise; e.g. assessments regarding other endpoints deemed relevant (such as environmental quality, nutrition, animal welfare, or specific economic factors, etc.).

The *main elements* of the evaluation process can be described as follows:

- Summarising the results of the assessment process – in terms of the likely consequences for human health or other relevant endpoints – and the concerns that individuals, groups or different cultures may attribute to a given food safety problem, both under the condition that no management measures were taken;
- Deliberation over these results in consideration of wider social and economic factors (e.g. benefits, societal needs, quality of life factors, sustainability, distribution of risks and benefits, social mobilization, and conflict potential), legal requirements, and policy imperatives;
- Weighing pros and cons and trading-off different – sometimes competing or even conflicting – preferences, interests, and values with regard to a given threat; or a trade-off analysis of a set of functional equivalents of the substance, product, process, or practice under consideration (the framework envisions such a broader trade-off analysis under the condition of scientific uncertainty and as the step following a precautionary assessment);
- Conclusion on whether the given threat is acceptable, tolerable, unacceptable, or ill-defined, or on what is the most appropriate functional equivalent. Should the threat be ill-defined, the assessment process needs repeating or augmenting;
- If management measures are deemed necessary, the most appropriate management approach should be recommended (details of which will be discussed in Sect. 5.4).

The term *tolerable* refers to an activity that is seen as worth pursuing (for the benefit it carries), yet requiring additional efforts for threat reduction within reasonable limits. The term *acceptable* refers to an activity where the remaining threats are so low that additional efforts for threat reduction are not seen as necessary. If tolerability and acceptability are located in a threat diagram – with probabilities on the *Y*-axis and extent of consequences on the *X*-axis – the well-known "traffic-light model" emerges. In this grey scale variant of the model (Fig. 5.2) black stands for intolerable threat, dark grey indicates tolerable threat in need of further intervention actions, and light grey shows acceptable or even negligible threat. The spotted area illustrates the borderlines: the first border identifying the area approaching certainty (probability = 1), and the second, where one gets close to indefinite losses. In both cases, the framework suggested here would recommend preventive actions.

To draw the line between "intolerable" and "tolerable" as well as "tolerable" and "acceptable" is one of the most difficult tasks of safety governance.[1] Yet such a

[1] The UK Health and Safety Executive (HSE) developed an evaluation procedure for chemical risks based on risk-risk comparisons (cp. Löfstedt 1997). Some Swiss cantons such as Basle County experimented with Round Tables as a means to reach consensus on drawing the two lines, whereby participants in the Round Table represented industry, administrators, county officials, environmentalists, and neighbourhood groups (cp. RISKO 2000: 2–3). Irrespective of the selected means to support this task, the judgement on tolerability or acceptability is contingent on making use of a variety of different knowledge sources.

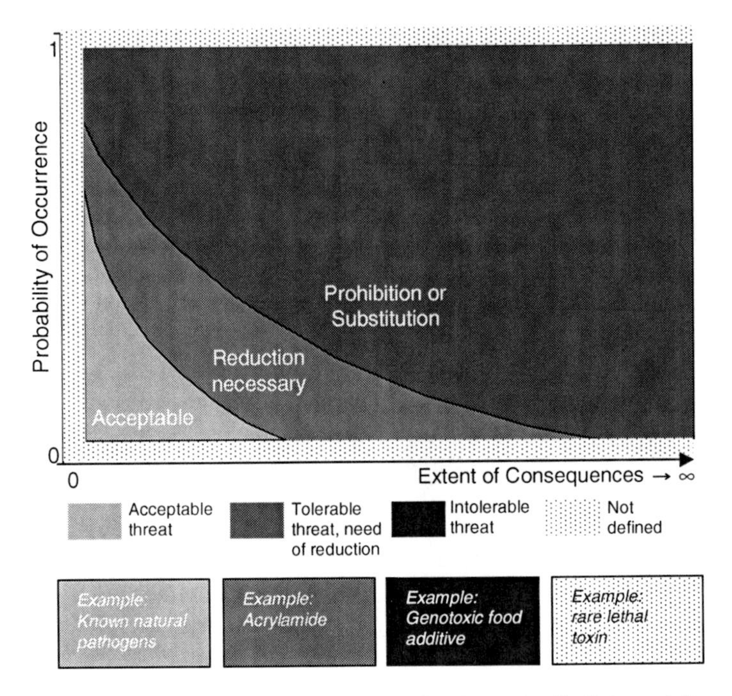

Fig. 5.2 Acceptable, tolerable, intolerable and borderline threats (traffic-light model)

judgement is required in order to proceed with decisions on management requirements. Arriving at a *balanced judgement* means that the assessed product, process or technology will render sustainable added value for society, economy, and industry only if the associated threats may be controlled and managed in a way acceptable to society. It does not suffice to include the "physical-risk" approach only – despite its undoubted importance – as it addresses but part of what is at stake within culturally plural, morally concerned and educated societies (Grove-White et al. 2000). Stakeholders play an important role in defining what is acceptable or intolerable by considering, among other things, the balance between risk and benefits and the probability of extreme events. Therefore the General Framework proposes to involve them as formal members of the Interface Committee, the proposed body with the mandate to give *advice* to the European Commission with regard to evaluation decisions, and/or to involve them through the Internet Forum (the baseline for a food safety interface structure) where stakeholders would be invited to deliberate on the evaluation advice or decision (see Chap. 6 for a detailed discussion of these options proposed).

After the evaluation exercise has been conducted by the Interface Committee or the European Commission (if no Interface Committee were to be set up), management is being presented with three potential outcomes:

- *Intolerable situation:* this means that either the threat source (such as a technology or a chemical) must be abandoned or replaced or, in cases where this is not possible, vulnerabilities need to be reduced and exposure restricted.
- *Tolerable situation:* this means that the threats must be reduced or handled in some other way within the limits of reasonable resource investments – "As Low As Reasonably Practicable" (ALARP) – (including best practice). This can be done by private actors (such as corporate risk managers), or public actors (such as regulatory agencies), or both (public–private partnerships).
- *Acceptable situation:* this means that the threats are so small – perhaps even regarded as negligible – that any threat reduction effort is unnecessary. However, threat sharing via insurances and/or further threat reduction on a voluntary basis, present options for action which can be worthwhile pursuing even in the case of an acceptable threat.

The distinction in intolerable, tolerable, and acceptable may appear (too) simple but it reflects the actual need for a judgement at the end of the assessment and evaluation processes. This final judgement on the given food safety problem allows for only three alternatives: either to do nothing, to ban the threat, or to initiate threat-modifying actions. There is no other alternative at this point. The governance framework – as presented here – emphasises that this important judgement is to be made as *transparent* as possible to all interested individuals and parties and that the institutions responsible for this judgement have the skills, the assets, the background knowledge, and the sensitivity with respect to the corresponding values and socio-cultural preferences to arrive at an informed, balanced, and fair judgement.

With regard to the three evaluation outcomes, the managers may either face a situation of unanimity, i.e. all relevant actors agree with how a given threat should be qualified, or a situation of conflict in which major actors challenge the classification made by others. The *degree of controversy* is one of the drivers for selecting the appropriate instruments for the type of *participation procedure* needed to resolve these controversies. The use of additional participation processes which reach beyond the inclusion of stakeholders through the respective food safety interface institution(s) (i.e. the Internet Forum and the Interface Committee) will depend on the case in hand and be considered by the Interface Committee or, if this interface institution is not established, by the Commission solely. The *prima facie* default is as follows: If there is hardly any ambiguity and controversy, participation and deliberation through the Interface Committee and/or the Internet Forum are likely to be sufficient as a means of eliciting the evaluation criteria, risk-benefit ratios, and trade-offs of a diversity of social groups. If the topic raises strong controversy and evaluation is highly ambiguous, a full-fledged participation process might be appropriate. Deliberation through the Internet Forum could be complemented by face-to-face participatory deliberation processes such as stakeholder roundtables, citizen forums, citizen juries or consensus conferences. In this situation, citizens' face-to-face deliberation could be part of the exercise in processes, where a randomised or deliberately stratified group of individuals work to scope and explore the issues and options in contention (see Chap. 7 for a detailed discussion).

5.3 Decision-Making in Management

As in conventional understandings of the governance of food safety, the final major stage in the General Framework is management. As a part of this framework, it has essentially the same meaning as the definition given in the General Food Law (Art. 3 (12)) and is therefore conducted by both the Commission and the Member States. It starts with a review of all the relevant information gained in the assessment process and the tolerability/acceptability judgement and the recommendation for the most appropriate management approach with which the evaluation exercise concluded. On that basis management measures are identified, selected, and implemented.

Hence, it is at this point of the governance cycle that *decisions* on management measures are being taken. This requires the consideration of policy choices among contending possible management measures. Such measures may include numerical limits for concentrations of substances in food items, standards for production and consumption, performance control, food preparation guidelines, monetary incentives, labels, and others. In some ways, this is analogous to the process already undertaken in assessment and evaluation. Here, however, the information is based on the positive and negative implications of a series of different intervention measures and not of particular threats (i.e. specific substances, products, processes, or practices). Depending on the context, the relevant information might best be gathered through the terms of reference for assessment itself, by reference to the most relevant measures. In other cases, it will be necessary to undertake this information-gathering process at the management stage in addition – and as a complement – to the evidence gathered during the assessment. Either way, the series of steps involved in the decision-making process on management measures is as follows (IRGC 2005: 40–48):

(1) *Identification of possible management measures (under special consideration of the suggestions made during the evaluation stage):* Generic management measures include the avoidance, the reduction and the transfer of a given threat and – also a measure to take into account – restraint. Whereas to avoid a threat means either selecting a path which prevents exposure (e.g. by abandoning the development of a specific technology) or taking action in order to fully eliminate a certain threat, threat transfer deals with ways of passing the threat in question on to a third party. Restraint as a management measure essentially means taking an informed decision to do nothing about the threat and to take full responsibility both for the decision and any consequences occurring thereafter. Management by means of threat reduction can be accomplished by many different means. Among them are:

- Technical standards and limits that prescribe the permissible threshold of concentrations, the take-up or other measures of exposure;
- Performance standards for technological and chemical processes;
- Governmental economic incentives including taxation, duties, subsidies, and certification schemes;
- Third-party incentives, i.e. private monetary or in-kind incentives;
- Compensation schemes (monetary or in kind);

- Insurance and liability;
- Cooperative and informative measures ranging from voluntary agreements to labelling and education programs.

All these measures can be used individually or in combination to accomplish even more effective threat reduction. Measures for threat reduction can be initiated by private and public actors or both together.

(2) *Assessment of management measures (with respect to predefined criteria):* Each of the measures will have desired and unintended consequences which relate to the threats they are supposed to reduce. In most instances, an assessment should be made according to the following criteria:

- *Effectiveness:* Does the measure achieve the desired effect?
- *Efficiency:* Does the measure achieve the desired effect with the least resource consumption possible?
- *Minimisation of external side effects:* Does the measure infringe on other valuable goods, benefits or services such as competitiveness, public health, environmental quality, social cohesion, etc.? Does it impair the efficiency and acceptance of the governance system itself?
- *Sustainability:* Does the measure contribute to the overall goal of sustainability? Does it assist in sustaining vital ecological functions, economic prosperity, and social cohesion?
- *Fairness:* Does the measure burden the subjects of regulation in a fair and equitable manner?
- *Political and legal implementability:* Is the measure compatible with legal requirements and political programmes?
- *Ethical acceptability:* Is the measure morally acceptable?
- *Public acceptance:* Will the measure be accepted by those individuals who are affected by it? Are there cultural preferences or symbolic connotations that have a strong influence on how the threats are perceived?

(3) *Evaluation of management measures:* This step integrates the evidence on how the measures perform with regard to the assessment criteria with a value judgement about the relative weight each criterion should be assigned. Ideally, the evidence should come from experts, and the relative weights from politically legitimate decision makers including stakeholder input. In practical management, the evaluation of measures should be done in close cooperation between experts and decision makers.

(4) *Selection of one or more appropriate management measures:* Once the different measures are evaluated, a decision has to be made as to which measures are to be selected and which rejected. This decision is obvious if one or more measures turn out to be dominant (relatively better on all criteria). Otherwise, tradeoffs that need legitimisation will have to be made (Graham & Wiener 1995). A legitimate decision can be made on the basis of formal balancing tools (such as cost–benefit or multi-criteria-decision analysis), by the respective decision makers (provided this decision is informed by a holistic view of the problem) or in conjunction with participatory procedures.

In the broader understanding of management, this stage involves two more steps:

(5) *Implementation of management measures:* It is the task of management to oversee and control the implementation process. In many instances implementation is delegated, as when governments take decisions but leave their implementation to other public or private bodies or to the general public. However, the management team has at any rate the implicit mandate to supervise the implementation process or, at least, monitor its outcome.
(6) *Monitoring how these measures perform in practice:* The last step refers to the systematic observation of the effects once the measures have been implemented. The monitoring system should be designed to assess intended as well as unintended consequences. Often a formal policy assessment study is issued in order to explore the consequences of a given set of management measures on different dimensions of what human beings value. In addition to generating feedback for the effectiveness of the measures taken to reduce the threats, the monitoring phase should also provide new information on early warning signals for both new and old threats viewed from a different perspective. It is advisable to have those responsible for performing the risk and concern assessments and the precautionary assessment, participate in monitoring and supervision so that their analytic skills and experience can be utilised in evaluating the performance of the selected management measures.

These steps follow a logical sequence but can be arranged in different orders depending on both situation and circumstance. It might be helpful to visualise the steps not as a linear progression but as a circle forming an *iterative process* in which reassessment phases are intertwined with new measures emerging, new situations arising or new demands being placed on managers. Similarly, sometimes the assessment of different measures causes the need for new measures to be created in order to achieve the desired results. In other cases, the monitoring of existing rules impacts on the decision to add new criteria to the portfolio. Measure identification, information processing, and measure selection should indeed be seen as a dynamic process with many iterative loops.

Table 5.1 provides a summary of the management steps. The list of examples and indicators represents the most frequently used heuristic rules for selecting input and for measuring performance.

5.4 Approaches to Management

In analogy to assessment, the framework also distinguishes between four management approaches. These are prevention, a precaution-based approach, a concern-based approach, and a risk-based approach. Each of these approaches lends itself to a set of suitable management measures (as shown in Table 5.2). There is *no automatic correlation* in the allocation of assessment and management approaches, yet there is a *preliminary assumption* that the appropriate assessment approach is subsequently pursued during the phase of management.

Table 5.1 Generic management components

Management components	Definition	Examples/indicators
1 Identification	Identification of potential measures, in particular threat reduction, i.e. prevention, adaptation and mitigation, as well as threat avoidance, transfer and restraint	– Standards – Performance rules – Restrictions on exposure or vulnerability – Economic incentives – Compensation – Insurance and liability – Voluntary agreements – Labels – Information/education
2 Assessment	Investigations of impacts of each measure (economic, technical, social, political, cultural)	– Effectiveness – Efficiency – Minimisation of side effects – Sustainability – Fairness – Legal and political implementability – Ethical acceptability – Public acceptance
3 Evaluation and selection	Evaluation of measures (multi-criteria analysis) and decision taking	– Assignment of trade-offs – Incorporation of stakeholders and the public
4 Implementation	Realisation of the most preferred measure	– Institutional accountability – Organisational efficiency – Cost-effectiveness of implemented measures
5 Monitoring and feedback	- Observation of effects of implementation (link to early warning) - Ex-post evaluation	– Investigation of intended impacts – Investigation of non-intended impacts – Policy impacts

5.4.1 Prevention

This approach applies where threats have been identified in the assessment process as certainly and unambiguously to be serious. Existing preventive approaches yield a wide variety of instruments and measures appropriate for the reduction, phasing-out or banning of the activities or products in question. The only management objective here is to eliminate the threat-causing activity in a fashion that is as economically efficient and socially acceptable as possible. If the assessment process has brought to light any mitigating factors that justify conditional relaxation of restrictive regulatory instruments, evaluation may, however, address the possibility that the threat may nonetheless be tolerated if the benefits or justifications were sufficiently overwhelming.

Table 5.2 Four management approaches

Management approach	Suitable measures include:
Prevention	– Bans (substitution possible?) – Phase-outs (substitution possible?) – (tolerance only when benefit is overwhelming)
Precaution-based	– Containment in space and time[2] – Close monitoring of potentially adverse effects – (More) stringent provisions for compensation and liability – Selecting the functional equivalent with a significantly lower risk and/or less uncertainty – Bans (substitution possible?) – Phase-outs (substitution possible?)
Risk-based	– Technical standards – Economic incentives – Labelling and information – Voluntary agreements
Concern-based	All of the above: choice is highly dependent on the outcome of participatory procedures of stakeholder and public engagement

Whilst depending intrinsically on the case in question, the criterion of sufficiency must, however, itself be extremely rigorous. Subject to the governance principle of participation, such a criterion could only be determined and applied through a broad-based process of participatory deliberation which might include both Internet-based *and* face-to-face deliberation, and would need to be further legitimated through dedicated procedures of democratic accountability.

5.4.2 Precaution-Based Approach

A precaution-based approach is required under the condition of unresolved scientific uncertainties. These imply that the (true) dimensions of the threats are not (yet) known. Therefore, it is vital to pursue a cautious strategy that allows learning by restricted errors. This management strategy needs to be informed by processes of precautionary assessment (detailed in Sect. 4.3) and a trade-off analysis of a set of functional equivalents of the product, process, or practice under consideration performed at the stage of evaluation. This trade-off analysis requires a more extensive

[2] The containment approach allows small steps in implementation enabling the managers to stop, or even reverse, the process as new knowledge is being produced or the negative side effects become visible. It is applied in European regulation of GM crops. Principally, for each case a risk assessment is carried out and the likelihoods of characterized hazards are determined by successively larger-scale experiments (case-by-case, step-by-step approach).

reflection on and deliberation over the effects that the different choices would imply in different dimensions. The consideration of wider social and economic factors is here of particular relevance (as in the case of high degrees of socio-political ambiguity, see the concern-based approach), as is the resort to "trans-disciplinary" deliberation involving specialists from ethics, humanities and social (as well as natural) sciences, alongside active engagement by a diversity of interested and affected parties through the Internet Forum, and possibly also through *face-to-face* participatory deliberation processes (for more detail see Chap. 7). Specifically, precautionary management measures may include, for example, small steps in implementation (containment approach) and close monitoring of potential side effects that enable managers to stop or even reverse the process as new knowledge is being produced or the negative side effects become visible. They may also be associated with enhancing the resilience of threat-bearing systems so they can better cope with surprises. Strategic options for resilience include diversification of the means for approaching identical or similar ends and reducing overall catastrophic potential or vulnerability. They may further include an emphasis on the substitution of those products, processes or technologies presenting the greatest threats and more stringent provisions for compensation, including strict and absolute liability regimes, mandatory insurance requirements, and product-withdrawal schemes.

5.4.3 Risk-Based Approach

For those threats, which can be adequately described by the two classic components "probability" and "extent of harm" (on the basis of more or less sophisticated data modelling depending on the complexity of the given threat), management measures may include, for example, the setting of technical standards, economic incentives, education, labelling and voluntary agreements. Measures to deal with more complex risks where it is more difficult to establish the cause-effect relationship between the risk agent and its potential consequences, may further include additional safety factors or redundancy and diversity in the design of safety devices. Evaluation can be done on the basis of traditional methods such as risk–risk comparison (for instance, does the new activity replace an established activity with a greater risk to human health, or would an established activity be substituted by an activity implying a greater risk to human health?), cost-effectiveness and cost–benefit analysis or balancing of risks and benefits with a clear priority on human health effects. Certainly, the proper use of these instruments requires transparency over subjective "framing assumptions", sensitivities and limits to applicability and their implications for the shaping of parameters on both sides of the cost–benefit equation. Participatory processes beyond the Interface Committee and/or the Internet Forum at the stages of evaluation and management would not be required.

5.4.4 Concern-Based Approach

This strategy applies to situations in which there is intense controversy among key stakeholders and also different parts of the affected and/or observing wider public over the framing of the food safety problem, the appropriate ways to interpret the assessment results, and/or the need and requirements for management. The stakeholders on the Interface Committee, the concern assessment, and also the deliberations via the Internet Forum, are major sources of information about whether these conditions are given: Are there strongly divergent viewpoints on the type of problem given, the relevance, meaning and implications of factual explanations and predictions for deciding about the tolerability or acceptability of a given threat, and the values and priorities of what should be protected? As pointed out above, in such circumstances of high socio-political ambiguity there is the need to organise a *broad societal discourse* in which issues of fairness, visions of future technological developments and societal change, and preferences about desirable lifestyles and community life play a major role, preferably at the stage of evaluation. Compared with the situation of scientific uncertainty, it is of even higher relevance that under this condition management is informed by the conclusions of a broad "trans-disciplinary" deliberation at the evaluation stage. As will be described in more detail in Chaps. 6 and 7, the Internet Forum is a means of generally assuring that all stakeholders and also representatives of the wider public can question and collectively consider all major elements of the governance process, including evaluation and management decisions. When food safety problems are subject to strongly divergent cultural attitudes, political perspectives, or economic interests, it might be required to, in addition, organise *face-to-face* participatory deliberation processes involving all relevant stakeholders and/or representatives of the wider public. If the choice of the appropriate management measures is highly contested as well, both stages, evaluation and management, might need to be subjected to extended participation. Applicable methods include randomly selected citizens' panels or juries, voluntary advisory groups, consensus conferences, and other face-to-face participatory techniques aimed at resolving ambiguities and value conflicts. The aim of this more extensive participatory deliberation is to "close down" on the most robust basis for consensus or common ground in decision making (informed by processes of concern assessment which, as outlined above, "open up" the salient features of the ambiguities in question and the particular divergences of perspective; see Chap. 7 for a more detailed discussion). At the end, in management, discrete measures need to be selected and implemented.

Following this approach to management, the intervention measures to be adopted may include any of those listed above as appropriate to prevention, precaution, or risk-based approaches. The significant difference with the concern-based approach is that measures will be highly dependent on the outcome of procedures of stakeholder and public engagement.

Chapter 6
Legal and Institutional Aspects of the General Framework

E. Vos and F. Wendler

6.1 Introduction

As stated in previous chapters, one of the primary objectives of the General Framework is to be fully compatible with the existing legal requirements of EU food safety regulation and to be implementable with as few institutional changes as possible. Following this objective, it is stressed at the outset that the General Framework could be put into practice without any major structural changes within the current system, by taking into account its procedural and methodological recommendations. This applies especially to the handling of different types of food safety threats, the involvement of stakeholders, and an increased awareness of the need for a transparent and consistent coordination between assessment and management, in particular with regard to the tasks of framing and evaluation. Yet, it is argued that some limited institutional changes would *facilitate* the realisation of the innovative steps of food safety governance established by the General Framework, especially the tasks of screening, the setting of terms of reference and evaluation, and the reconsideration of participation procedures. Therefore, this chapter is aimed at setting out a proposal for such limited institutional changes. These recommendations are summarised in Table 6.1 below, together with the legal and institutional issues involved.

6.2 Proposal for Institutional Changes

This proposal for limited institutional changes as recommended by the General Framework consists mainly of three parts:

(1) The creation of a *Screening Unit* and a *Panel on Concern Assessment* within EFSA as part of a proposal for the improvement of the capacities of EFSA to fulfil the functions foreseen in the General Framework;
(2) The establishment of *food safety interface institutions* to improve the inclusiveness, transparency and coherence of the setting of terms of reference and evaluation;

M. Dreyer and O. Renn (eds.), *Food Safety Governance*,
DOI: 10.1007/978-3-540-69309-3_7,

Table 6.1 Proposal for limited institutional changes

	Main functions within the general framework	*Embeddedness in existing structures of food safety regulation and the General Food Law (GFL)*	*Responsible actor(s) in current arrangements*	*Institutional changes recommended?*
Framing: review	General development and oversight of food safety governance	Variety of arrangements and procedures	EP, Council, Commission, EFSA, Member States (Standing Committee on Food Chain and Animal Health)	No specific changes; however, comments made through the internet forum and the interface committee may lead to review, thus leading to modification of existing procedures (e.g. existing screening criteria)
Framing: referral	Identifying problem and applicable legislation and referring the matter to EFSA for screening	Consultation of EFSA as required by Art. 29 GFL or specific procedures; e.g. according to Reg. 1829/2003 EC; Dir. 2001/18 EC et al.	Commission, EP, Member States	No specific changes; however, comments made through the internet forum may lead to referral, thus leading to modification of existing procedures
Framing: terms of reference	Specification of the terms of reference tor the assessment	Setting the terms of reference by the originator of the request for an opinion; co-ordination by DG SANCO when opinion is requested by the Commission	Commission, EP or Member States	Yes: creation of an interface committee (with two options proposed), and an internet forum
Screening	Identification and basic characterisation of threats	Definition of risk assessment in Art. 3 (11) GFL, description of tasks of EFSA in Art. 22 (4) and 23 (f) GFL	EFSA	Yes: creation of Screening Unit within EFSA
Assessment	Applying the four approaches to assessment identified in the general framework	Definition of risk assessment in Art. 3 (11) GFL and related articles	EFSA	Yes: creation of a Concern Assessment Panel within EFSA; better application of existing rules: re-consideration of participation procedures
Evaluation	Conducting a tolerability/acceptability judgement	Definition of risk management in Art. 3 (12) GFL, consideration of "other legitimate factors" in authorisation procedures (Art. 7, Reg. 1829/2003 EC)	Commission, Member States	Yes: creation of an interface committee (with two options proposed), and an internet forum
Management	Identify, assess, select and implement measures	Definition of risk management in Art. 3 (12) GFL and related articles	Commission, Member States	Better application of existing rules: reconsideration of voting requirements in comitology procedure, and participation procedures

(3) Better application of *existing rules*, both with regard to stakeholder involvement and decision-making procedures in the framework of the comitology procedure.

With the proposed institutional changes, the General Framework seeks to implement the following objectives and principles. It is aimed at:

(a) Introducing more *transparency* into the conduct of food safety governance procedures, in particular the drafting of terms of reference, evaluation, and decision-making at the stage of comitology. This objective builds on efforts made by both EFSA and the Commission to increase the transparency of risk assessment and to achieve a better understanding of the limitations of science by risk managers, key stakeholders, and the public.[1] Furthermore, this objective builds on calls by the Commission for transparent decision-making procedures regarding the application of the precautionary principle in this area (CEC 2000a: 18);
(b) Achieving better *involvement* of stakeholder organisations and the wider public, particularly in relation to the scientific uncertainty and socio-political ambiguity involved in a given food safety threat. It takes as basis both the Commission's White Paper on European Governance (CEC 2001a) stressing the importance of the principle of participation, and its Communication on the Precautionary Principle, urging for the involvement of all interested parties in the decision-making process at the earliest possible stage (CEC 2001a);
(c) Ensuring the *effectiveness* and *flexibility* of procedures of food safety governance and avoiding bureaucratic overload, a need very much highlighted by policy practitioners in the workshop-based feedback and review process (see Chap. 11, this volume). The general principle of effectiveness is also enshrined in the Commission's White Paper on European Governance;
(d) Embedding the innovative procedures of framing (review, setting the terms of reference), screening and evaluation as far as possible within the *existing structures*, in order to make the General Framework as easily applicable as possible and reduce the costs of institutional innovations to a minimum;
(e) Providing for procedures for handling threats which involve scientific uncertainty and socio-political ambiguity that comply with both legal requirements and principles at European level and *international agreements* in the framework of the WTO (in particular the WTO's Agreement on the Application of Sanitary and Phytosanitary (SPS) Measures), as these represent fundamental obligations on the conduct of food safety regulation at European level.

The proposed institutional changes mentioned above – in relation to the improvement of capacities of EFSA and the design of the food safety interface institutions – are outlined in more detail in the following sections.

[1] As expressed in various working documents of both institutions already discussed in Chap. 1, this volume.

6.3 Improved Capacities of EFSA

The General Framework recognises EFSA as the central actor for risk assessment and does therefore not alter the distribution of tasks between EFSA and the Commission established by the General Food Law (GFL).[2] However, it proposes two limited innovations with the objective of increasing the capacity of EFSA to take on the functions allocated to it by the General Framework. These recommendations relate to the conduct of screening by means of a specifically designed unit and the creation of a panel for concern assessment, which are set out below.

6.3.1 *Screening*

The tasks of hazard identification and characterisation undertaken through screening are part of risk assessment as defined in the General Food Law (GFL, Art. 3 (11)), and should therefore be fulfilled by EFSA. This is underlined by the enumeration of the tasks of EFSA in the GFL, which include the duty to "collect and analyse data to allow the *characterisation* and monitoring of risks which have a direct or indirect impact on food safety" (Art. 22 (4)) [present authors' emphasis]. Similarly, the GFL establishes the task of EFSA to "undertake action to *identify* and *characterise* emerging risks, in the field within its mission" (Art. 23 (f)) [present authors' emphasis].

In the present situation, it appears that, while EFSA may be scientifically equipped to undertake the task of screening, there is currently no specific department or unit capable of co-ordinating the referral of screening questions to the Scientific Panels and expert services.[3] Therefore, it is recommended that a new structure should be created with a specific responsibility for the conduct of this task, acting as a coordination point for the referral of questions and the collection of the corresponding answers from the responsible scientific units. This implies the creation of a "Screening Unit", established as a small structure which would mainly have the task of acting as secretariat for the conduct of screening. It would therefore *not* conduct the investigation of the questions asked through screening *itself*, but have the task of passing on requests for screening to the different scientific panels or EFSA's Scientific Expert Services (such as in the fields of data collection, pesticides, zoonoses, and further units that are currently being established), and potentially also the various Working Groups established under the auspices of many Scientific Panels.

---

[2]European Parliament and Council Regulation (EC) No 178/2002 (*OJ* 2002, L31/1) as amended by Regulation (EC) No 1642/2003 (*OJ* 2003, L 245/4), hereinafter referred to as the *General Food Law* (GFL).

[3]These insights were gained both at the SAFE FOODS subproject 5 workshop with risk managers (Fondation Universitaire, Brussels, 23/24 October 2006; see Vos & Wendler 2006c) and the workshop with industry representatives (Haigerloch Castle, 18/19 September 2006; see Dreyer et al. 2006b).

Within the organisational structure of EFSA[4], the Screening Unit would be inserted as part of the department on "risk assessment" and would be conceived as a structure co-ordinating and overseeing the permanent scientific units of EFSA, which exist with parallel mandates to the Scientific Panels.

6.3.2 Concern Assessment

With regard to concern assessment, it was noted in the discussions at the workshops with key actors in food safety governance (cp. Chap. 11, this volume) that EFSA lacks the social scientific expertise to undertake this task, and that concern assessment could best be undertaken through a specific structure that works in parallel, and in interaction with the existing Scientific Panels. The General Framework thus envisions the creation of a "Concern Assessment Panel" to serve EFSA, in combination with a specific unit with social scientific expertise within EFSA's scientific expert services. The creation of a new Scientific Panel would require a decision by the Commission in the framework of the comitology procedure, made at the request of EFSA (Articles 28 (4) and 58 (2) of the General Food Law). The last adjustment of EFSA's Scientific Panels was undertaken through Commission Regulation 575/2006 (CEC 2006), which added the Panel of Plant Health (PLH). The creation of an additional unit of the scientific expert services would require action by the Management Board of EFSA which has the task of ensuring that EFSA carries out its mission and the tasks assigned to it (GFL, Art. 25 (7)).

6.4 Interface Between Assessment and Management

6.4.1 The Need for More Transparency, Participation and Coordination

The feedback gathered from the series of workshops highlights the need to avoid an overburdening of procedures and the addition of unnecessary bureaucratic layers in the attempt to introduce more transparency, participation and co-ordination.[5] The General Framework supports that position. In advocating more transparency and

[4]For an organigramme of EFSA, see: http://www.efsa.europa.eu/etc/medialib/efsa/about_efsa/structure/ 126.Par.0003.File.dat/comm_efsaorganigr_en.pdf. Accessed: 30 May 2007.

[5]In an earlier version of the General Framework (Stirling et al. 2006) we had proposed both the establishment of an "Operational Committee" (suggested in two slightly different forms) to be composed of assessors, managers and stakeholder representatives with the task of discussing terms of reference and evaluation, and a more flexible ad-hoc consultation procedure under the auspices of the Commission. Two primary problematic issues were brought up during the workshops, (1) the bureaucratic overload of the proposed institutional and procedural changes, and (2) the difficulty or impossibility to select representatives of stakeholders to be present on the "Operational Committee" (these criticisms are dealt with in more detail in Chap. 11, this volume).

participation and improved co-ordination in the interface procedures, the General Framework addresses the current problems in the risk governance process as well as the answers to various proposals by both EFSA and the Commission to increase the transparency of risk assessment. An EFSA guidance document to the Advisory Forum on increasing the transparency of risk assessment calls for a close information exchange between the EFSA Scientific Committee or Panel and the originator of a request for a scientific opinion, recognising that while the General Food Law

> provides for a clear distinction between risk assessment and risk management, … an efficient and transparent mechanism of interaction is obviously needed to ensure that appropriate exchanges may satisfactorily take place, particularly in more complex cases (EFSA 2006a: 9).[6]

The guidance document furthermore states that these interactions should seek to ensure that the terms of reference of questions put to EFSA are clearly drafted, and that opinions provided by EFSA are clearly formulated with the underlying science, indicating uncertainties in the assessment, so that "the information given in the opinion can be well understood and used by the originator of the request" (EFSA 2006a: 9). This guidance document follows an earlier information note by EFSA on increasing the transparency of risk assessment (EFSA 2004a: 5). In this document, EFSA expresses its plans to shed more light on the terms of reference and to include in it a description of the strengths and limitations of the data used and the underlying assumptions, the criteria for inclusion or exclusion of available scientific information for a given risk assessment, considerations about appropriate stakeholder engagement and other process-related issues, consistent documentation, and science-based statements about the need of additional studies for the conduct of a risk assessment. Similarly, as already mentioned, DG SANCO has also highlighted the need for good interaction and communication between risk assessors and risk managers, and suggested a formal procedure through which scientific groups in charge of risk assessment should designate two representatives to meet risk managers before the start of an assessment and again after the establishment of a draft scientific opinion (DG SANCO 2005; Vos & Wendler 2006b: 121).

6.4.2 Institutionalising Food Safety Interfaces

A large part of the innovative proposals set out in the General Framework refers to the *interface* between the spheres of assessing and managing food safety threats. First, this interface comprises the task of setting the terms of reference. As pointed out through our empirical research, strong interactions between risk managers and risk assessors can be observed at this stage within the current institutional framework

[6]The relevant passage of the document reads as follows: "A clear formulation of the question (i.e. "terms of reference") is another important step before carrying out any risk assessment. These "terms of reference" should include a clear definition of the concern and a plan for characterising and assessing the risk. Ideally, formulation of the "terms of reference" should be considered as an iterative process involving dialogue with stakeholders, where appropriate."

of EU food safety regulation. In current practice, the terms of reference are specified by the institution or authority that requests an opinion (Commission, Parliament or Member State). Currently, that is mostly the Commission. A specific unit of DG SANCO deals with the relations with science and stakeholders and is always involved whenever terms of reference are drafted and submitted. It co-ordinates all requests to EFSA for scientific opinions. This unit examines all mandates as to their background, tries to understand the type of answer the mandates are looking for, ensures the coherence with the other questions, sets the priority of the questions to be asked and establishes the legal basis under which to act. Here the exact phrasing of the question is spelled out, on the basis of the drafts made by the Commission officials dealing with the specific dossiers. The unit also functions as a "watchdog" in that it is charged with ensuring that Commission officials who attend meetings of the Scientific Panels of EFSA do not transgress their role as observers (Vos & Wendler 2006b: 120). While DG SANCO ensures co-ordination between the process of risk assessment and risk management in this way, one of the shortcomings of the current practice appears to lie in the fact that this is often done in a rather opaque manner, leading to calls by the Commission for more transparent communication and interaction (Vos & Wendler 2006b: 121). The General Framework thus aims to make this part of the "interface" more transparent and more inclusive.

Second, the interface relates to the step of evaluation which refers mainly to the consideration of the results of "risk assessment and other legitimate factors" relevant to the matter under consideration which are defined as a part of risk management by the General Food Law (Art. 3 (12)). Here again, our empirical research revealed that evaluation is currently part of risk management and, as such, is often done in a rather opaque manner, and that there is no systematic involvement of stakeholders. In order to avoid an overburdening with new structures, the General Framework, here too, considers that evaluation should be conducted in the *same structure* as the setting of terms of reference, thus involving actors from assessment and management as well as stakeholder organisations within the framework of the Interface Committee. Against this background, the General Framework seeks to establish an innovative structure in order to achieve a *more inclusive, transparent and systematic* co-ordination between assessment and management activities.

The General Framework recommends creating *food safety interface institutions* to improve both the transparency and consistency of the interaction between assessors and managers and the involvement of stakeholders herein. In this way it advocates a participatory process which goes beyond mere consultation and allows for more genuine *engagement.*

The General Framework thus proposes:

(a) To create an *Internet Forum* in order to increase the transparency of interface communication and documentation, and to allow for the broader engagement of stakeholders and the public with these communications; and
(b) Optionally, to create an *Interface Committee*, either in the form of a flexible and non-binding "Interface *Advisory* Committee" or in the form of a more compulsory and binding "Interface *Steering* Committee".

Whilst the General Framework unconditionally recommends the establishment of the Internet Forum as a means for increasing the transparency and openness of interface communications to stakeholders and the wider public, it offers two variants to formalise direct, face-to-face debates between assessors and managers in an institutionalised structure. The *three options* thus proposed are as follows (see Table 6.2):

(1) The *minimum* option, consisting only of the creation of the *Internet Forum* and leaving the direct interaction between assessors and managers to current practice without any further formalisation;
(2) The *maximum* option, consisting of the creation of the *Internet Forum* combined with the compulsory discussion of all cases in an *Interface Steering Committee* (ISC); or
(3) The *intermediate* option, consisting of the creation of the *Internet Forum* combined with a more flexibly applicable *Interface Advisory Committee* (IAC). As this option takes into account both the objective of establishing a more formalised setting for the interaction of assessors and managers, and the wish to keep the innovative structures sufficiently flexible, this option is proposed as the *preferred option* for the implementation of the General Framework.

When reflecting upon these proposals, it is important to underline that the proposal to introduce a more coherent and transparent step of setting the terms of reference, as expressed in options 2 and 3, coincides with the ideas by both the Commission and EFSA to work towards a more transparent and inclusive approach to defining the terms of reference in the process of risk assessment. Comparing the three options, we feel that the objectives of the General Framework are best expressed through an option that also includes a structure for the direct interaction between assessors and managers allowing for the necessary flexibility, which is suited best in the option of the Interface Advisory Committee.

Before explaining the practical operation of these options in more detail, we will first explain what we consider the composition and tasks of the Internet Forum and the Interface Committee in its two variants should be like.

6.4.2.1 Internet Forum

As explained above, the conduct of the interface tasks (terms of reference, evaluation) builds on the objective of eliciting the views of a wide range of assessors and managers at both European and national levels, stakeholders and the public. Against this background, the General Framework proposes to establish a web-based forum, which could work as a way to generally involve the *wider constituencies* rather than those being part of the Interface Committee: a wider diversity of civil society groups, but also risk managers and scientific experts, including the Member States' risk managers, and scientific experts affiliated to the Competent Authorities at national level. In this manner, the creation of the Internet Forum responds to the

Table 6.2 Three options for the institutional design of the food safety interface

Internet Forum only	Internet Forum and Interface Advisory Committee (IAC)	Internet Forum and Interface Steering Committee (ISC)
"Minimum option"	*"Intermediate (and preferred) option"*	*"Maximum option"*
	Interface Advisory Committee: *Composition:* – *Core members*: Equal number of assessors, managers, and stakeholder representatives, to be appointed by the Commission – *Case-specific members*: assessors, managers and stake-holders with specific expertise for different fields of food safety governance, to be appointed by the core members – *Ad hoc members*: may be invited by IAC for specific cases; e.g. experts or representatives of Member States (MS) or Europ. Parliament (EP)	*Interface Steering Committee:* *Composition:* – As in the IAC
	Tasks: – Gives advice on the terms of reference and evaluation to the Commission	*Tasks:* – Adopts terms of reference – Gives advice on evaluation to the Commission
	Working procedures: – Deals with but a selection of cases – IAC is convened by the Commission for particular cases, especially when screening has found uncertainty and/or ambiguity	*Working procedures:* – Deals with all cases of food safety governance

(continued)

Table 6.2 (continued)

Internet Forum only	Internet Forum and Interface Advisory Committee (IAC)	Internet Forum and Interface Steering Committee (ISC)
"Minimum option"	*"Intermediate (and preferred) option"*	*"Maximum option"*
Internet Forum *Framing:* – Publication of terms of reference – Exchange of views about referral and review *Assessment:* – Exchange of views on application of terms of reference in assessment procedures *Evaluation:* – Exchange of views on evaluation of food safety threats *Management:* – Exchange of views on management options	*Internet Forum* *(in addition to the tasks of the Internet Forum listed for the "minimum option"):* – Discusses proposals for the appointment of stakeholder representatives in the IAC – Makes suggestions for cases to be discussed by the IAC – Discusses evaluation results and the advice on terms of reference by the IAC	*Internet Forum* *(same tasks as in case of the IAC)*

concerns of inclusion, selection, and representation that unavoidably come up within the context of the membership of the Interface Committee.[7]

The General Framework therefore proposes to launch a website under the auspices of DG SANCO (managed by the Unit of Science and Stakeholder Relations), on which contributions made by (both European and national) assessment and management actors, stakeholders and also the wider public could be posted. It is envisaged that the Internet Forum should be organised in *four platforms*, relating to the main elements of the General Framework (Framing/Assessment/Evaluation/Management).

The Internet Forum would be used to increase transparency, especially through the publication of the draft terms of reference, but also to engage its participants in a debate and open exchange of views. With regard to the logic of involvement, the Internet Forum could serve, firstly, as a platform for both the targeted consultation of interest groups and civil society organisations by the Commission, EFSA and, possibly, the Interface Committee ("top-down"), and secondly, the more spontaneous and open elicitation of views and concerns of participants in the Forum ("bottom-up") (see Table 6.3 below).

(1) *Top-down*. In cases where specific responses of stakeholders and the public are sought with regard to particular cases of food safety governance, the Internet Forum could make use of involvement techniques with the *top-down* logic.

Table 6.3 Tasks and procedures of the platforms of the Internet Forum

Platforms	Tasks/procedures
Framing	– Publication and exchange of views on the (draft) terms of reference ("top-down")[8]; – Exchange of views and suggestions about referral and review ("bottom-up"); – Discussion of memberships in the Steering Committee or Advisory Committee if set up ("bottom-up").
Assessment	– Exchange of views on the results of screening ("top-down"); – Exchange of views on the application of terms of reference and assessment results ("top-down").
Evaluation	– Exchange of views on evaluation advice[9] and evaluation decisions ("top-down").
Management	– Consultation on proposals for management options ("top-down"); – Exchange of views on the choice of instruments and monitoring results ('bottom-up").

[7] As expressed by various participants in the workshops with key actors in food safety governance that we held between September and November 2006 (cp. Chap. 11, this volume).

[8] This point needs to be specified with regard to whether or not one of the Interface Committees has been set up, as these play a major role in the setting of terms of reference; see Table 6.2 providing an overview of the three options proposed below for further specifications.

[9] As in the setting of terms of reference, this task differs slightly when either the Steering Committee or the Advisory Committee has been established; see Sect. 6.4.2 which provides a discussion of the three options proposed for the food safety interface institutions for further specifications.

The debate in these cases would be rather strongly pre-structured (i.e. consultation documents would be posted on the website of the Forum with specific questions to be discussed), and the submission of comments would be restricted to a selected number of accredited stakeholder groups, to be chosen by the Commission.[10] This involvement technique could be applied to questions of a specific nature, such as the exchange of views about the draft terms of reference, screening results, the application of terms of reference and assessment results, and proposals for food safety management options. It is, however, stressed that the contributions of the participants should be directly visible on the website (and not just submitted to the Commission for consideration and summary), thus allowing for an exchange of views between the participants, and the evolution of genuine debate on the topics under discussion.

(2) *Bottom-up*. The *bottom-up* logic of involvement (i.e. one that follows the initiatives and concerns of interest groups, civil society organisations and the public) could be used to identify issues of concern to the widest possible variety of stakeholders and the public. This method could be applied to questions of a more general nature such as the discussion of the membership in the Interface Committee, the exchange of views on suggestions for referral and review (including the prioritisation of threats), and the exchange of views on the choice of management instruments and monitoring results. This implies that participants could take the initiative by suggesting which cases should be taken up for discussion, thus being able to make contributions to virtually any case of food safety governance. Importantly, these debates would not be mediated by the Commission or another institution, and hence contributions of participants would be posted on the website as they are, rather than being submitted to the Commission and then summarised in a report.[11] Furthermore, this kind of involvement would be open to all interested stakeholders and also the wider public. Given the potential problem of overcrowding, it is clear that this form of involvement could only be applied to a limited number of functions of the Internet Forum.

As mentioned above, the General Framework suggests that the Internet Forum could be *combined* with one of the two variants of the Interface Committee

[10] The stakeholders chosen in this context could include all members and associated members of the main stakeholder consultation bodies of EFSA and the Commission, the Stakeholder Consultative Platform and the Advisory Group on the Food Chain, and a selection of interest groups and civil society organisations to which contacts have been established by EFSA and by the Commission through specific consultations, e.g. the partners of the PRAPeR consultations on pesticides or organisations participating in debates on GMO with EFSA.

[11] Obviously, variations to these two points could easily be developed if this is desired (i.e. by restricting the range of cases that are up to discussion, and by introducing elements of summarising and mediation, i.e. through the establishment of contact points reporting from stakeholders and citizens to the website (as envisaged in existing Interactive Policy-Making initiatives of the Commission), or by introducing the existing technique of the Commission of gathering comments by e-mail and reporting them back to the website.

(i.e. either the Steering Committee or the Advisory Committee). Therefore, one of the main tasks of the Internet Forum would in this case be to communicate with, and comment on the work of the Interface Committee. In this context, it is stressed that the discussions in the Internet Forum would not directly determine the agenda of the Committee, but serve as an additional input to be considered by its members. However, not taking into account concerns expressed in the Internet Forum could lead to infringement of the principle of good administration, in particular the obligation for the Commission to examine carefully and impartially all the relevant elements of the individual cases.[12]

Therefore, the Internet Forum would involve a variety of involvement procedures in relation to the main tasks of interface communication, structured by the four platforms in relation to the four main governance stages envisioned by the General Framework. These include the following tasks shown in Table 6.3 below, indicating whether a procedure is applied with bottom-up or top-down logic.

6.4.2.2 Interface Committee

Apart from this web-based forum of involvement and debate, the General Framework also proposes a structure for the direct, face-to-face discussion between assessors, managers, and stakeholders. The two variants for this structure are being outlined in the following paragraphs.

(a) Interface Advisory Committee

A first possibility of ensuring the direct co-operation between those responsible for assessment and management would be to establish an *Interface Advisory Committee* (IAC) composed of assessors (i.e. members of EFSA Panels and scientific services), managers (i.e. members of units of DG SANCO in charge), and stakeholder representatives (including representatives of the key European consumer, industry and farmer organisations). The Committee would be established through a Commission Decision specifying its tasks and composition. The Interface Advisory Committee would adopt advisory opinions on the terms of reference of given cases and on the evaluation of cases addressed to the Commission. The institution or authority responsible for the definition of the terms of reference could then use these discussions to define the specific terms of reference forwarded to EFSA for an assessment. In this option the draft terms of reference would be published as

[12] According to a consistent line of case law of the Court of Justice, in cases where a Community institution has a wide discretion, it has to observe the procedural guarantees conferred to by the Community legal order. Those guarantees include in particular the obligation for the institution in charge to examine carefully and impartially all the relevant elements of the individual case. This will enable the Courts to ascertain whether the elements of fact and of law on which the exercise of the discretion depends were present. See e.g. Case C-269/90 *Technische Universität München* [1991] ECR I-5469, paragraph 14. We will leave outside the scope of this chapter the problematic issue of access to justice for individuals for Community acts.

soon as submitted to EFSA (in the current system, the terms of reference are revealed only after the completion of an opinion by EFSA).[13]

The Interface Advisory Committee would not be expected to deal with all cases of food safety governance, but to address only those cases considered to be particularly problematic or requiring further discussion between assessors, managers, and stakeholders. In practice, this would mean that the Commission convenes meetings of the Advisory Committee wherever deemed necessary, especially in cases where the results of screening have indicated sources of uncertainty or ambiguity. As indicated above, the Internet Forum would also have the opportunity to make suggestions about cases to be dealt with by the Advisory Committee. Furthermore, the Commission would be free to use the IAC as a forum for an exchange of views about questions of a more general nature. Therefore, it is envisaged as a structure that operates remotely from individual decision-making procedures in single cases of food safety governance. It would be dealing with but those cases that only assessors, managers, and stakeholders would like to discuss in the framework of the Committee. This way, the IAC would also be free to combine or "bundle up" cases in a manner that appears conducive to the effectiveness of procedures and the avoidance of overload.

It is envisaged that the Interface Advisory Committee would work in a flexible setting, with its composition depending on the case in question around a core of permanent members (see Table 6.4). To this end, the Commission would appoint a group of core members of the IAC consisting of an equal number (2–4) of assessors, managers, and stakeholder representatives (suggesting a size of the core group between 6 and 12 committee members). These should include members of the *horizontal* units of EFSA and the Commission (i.e. those units responsible for non case-specific issues as science and stakeholder relations, risk assessment, food law, and the food chain), and stakeholders with a background in the representation of the general interests of consumers, industry, farmers, and other interests involved in the food chain.

Furthermore, in order to be able to deal with cases from different fields, the IAC should be convened in diverse constellations for each major field of food safety governance (suggesting 6–9 different constellations). These constellations could be established in correspondence with the eight Scientific Panels of EFSA. Therefore, in addition to the core committee members, each constellation of the IAC should include an equal number (2–4) of assessors, managers and stakeholder representatives with *case-specific expertise*. These committee members would be appointed by the core committee members and could be recruited from the Scientific Panels and scientific expert services of EFSA, the units of DG SANCO in charge of a

[13] In current practice, the terms of reference are published on the website of EFSA after an opinion has been established; See: http://www.efsa.europa.eu/EFSA/ScientificOpinionPublicationReport/efsa_locale-1178620753812_ScientificOpinions.htm. Accessed 20 June 2008. Ongoing assessments are listed in the register of questions including documentation on the mandate of a risk assessment, without, however, stating the exact terms of reference of the risk assessment; see: http://registerofquestions.efsa.europa.eu/roqFrontend/questionsList.jsf?nocache=1216672853433. Accessed 20 June 2008.

Table 6.4 Size and composition of the Interface Advisory Committee

	Managers	Assessors	Stakeholder representatives
Core committee members	2–4 persons representing "horizontal" units of DG SANCO (e.g. on science and stakeholder relations, food law, food chain and labelling)	2–4 persons representing "horizontal" EFSA bodies (e.g. Scientific Committee, units on science and risk/concern assessment)	2–4 persons having their background in the representation of general interests of consumers, industry, farmers or other interests of the food chain
Case-specific committee members	2–4 persons representing case-specific units of DG SANCO (e.g. on pesticides, GMOs or animal health)	2–4 persons representing case-specific bodies of EFSA (e.g. members of the scientific panels or of the scientific services)	2–4 persons with a back- ground in the representation of case-specific stakeholder interests
	To be appointed by the core committee members for all major fields of food safety governance (i.e. 6–9 different constellations of case-specific committee members)	To be appointed by the core committee members for all major fields of food safety governance (i.e. 6–9 different constellations of case-specific committee members)	To be appointed by the core committee members for all major fields of food safety governance (i.e. 6–9 different constellations of case-specific committee members)
	Plus: may invite ad hoc members for particular cases when considered necessary	*Plus*: may invite ad hoc members for particular cases when considered necessary	*Plus*: may invite ad hoc members for particular cases when considered necessary

specific field of food safety governance (such as pesticides, genetically modified organisms, animal health, etc.), and from national food authorities as well as stakeholder representatives with a case-specific interest.

Moreover, for particularly problematic cases, the IAC would be allowed to invite additional experts with specific interests or expertise on an *ad hoc basis*, depending on the case in question. For example, if the originator of a request to EFSA is a Member State or the European Parliament, a representative of the respective institution should be invited to the committee session as an ad hoc member. Although no fixed number of participants is prescribed for the IAC, it is clear that it should remain a sufficiently small structure to work effectively and therefore not include too many participants. The size of the IAC could therefore vary between 6 and 24 and be kept flexible, with the objective of bringing together the assessors and managers with specific expertise and responsibility for a given field of food safety governance (see Table 6.4).

(a) Interface Steering Committee

A second, more strongly formalised variant of ensuring deliberations between assessors, managers and stakeholders, is the creation of the *Interface Steering Committee* (ISC). The ISC would have the same size and composition as the Interface Advisory Committee, and also serve as a platform where the terms of reference and evaluation are discussed between the three actor groups. However, contrary to the Interface Advisory Committee, the Interface Steering Committee would *adopt* the terms of reference instead of issuing only an advisory opinion. The tasks of the ISC with regard to evaluation, however, would still be restricted to the adoption of advisory opinions.[14] Furthermore, the tasks of the ISC could be defined as dealing with *all* cases of food safety governance, instead of a selection of only the more problematic cases. This could take into account the view that it is not only the deliberation about the terms of reference which requires an open exchange between assessors, managers and stakeholder, but also the more fundamental decision on the *selection* of critical cases. From this point of view, it could be argued that it would be less laborious to deal with all cases (albeit at different intensities) than to organise a meeting of the Interface Committee in each case where assessors or stakeholders flag up critical issues. Moreover, the decision-making process on the selection of critical or special issues would also be subjected to full transparency. Therefore, whereas this variant may appear as more burdensome at first sight, it clearly has its advantages in carrying forward the objectives of openness, transparency and stakeholder involvement in a very obvious manner. The establishment of the ISC (in combination with the Internet Forum) is proposed as the "maximum" option for the design of the food safety interface institutions in the General Framework.

[14] For further explanations on this point, see Sect. 6.6.6 in this chapter on the principle of the non-delegation of powers (Meroni doctrine).

6.5 Management: Re-Consideration of the Comitology Procedure

"Management" as a part of the General Framework presented here has essentially the same meaning as the definition given in the General Food Law (Art. 3 (12)). One of the main recommendations of the General Framework with regard to this step of food safety governance refers to the re-consideration of procedures for the involvement of stakeholder organisations. It is important to stress, however, that the participation procedures (both at the stages of management and assessment) should be implemented without institutional changes. In this way, it will make use of existing arrangements and procedures and does *not* foresee the creation of another consultation body or forum.

In addition, it should be highlighted that a particularly important and sensitive question in management refers to the *application of the comitology procedure* in the adoption of measures, such as the approval of authorisations. Whereas comitology committees were initially created to serve as a control mechanism for the fulfilment of implementation tasks by the Commission, in practice they mostly appear to work as a strong mechanism for deliberative decision-making, advancing consensus as part of a regulatory network with a strong role for the Commission, thus raising questions about the transparency, control and oversight of the committees themselves. It is therefore unsurprising that comitology has been subject to intensive debates both in the academic and practitioners' circles.[15] Hence it appears as one of the key institutional challenges in the field of risk management to ensure the compliance of comitology procedures with principles of good governance (especially transparency and accountability) while preserving this procedure as a pragmatic and powerful mechanism for deliberative decision-making and the creation of consensus around the adoption of measures in risk management.

Following the comitology (regulatory committee) procedure, the Commission may currently adopt implementing measures (including authorisations) notwithstanding the absence of a political agreement among the Member States. In view of the problems that this approach causes, the General Framework recommends that in areas such as the authorisation of genetically modified food products, implementation decisions by the Commission should not be adopted in the absence of a qualified majority vote expressing the political support of a majority of the Member States for the adoption of such a decision. This recommendation might possibly be made without the requirement for changes in the institutional framework, as it would actually follow existing commitments expressed by the European Community (EC) institutions. As the Commission does not seem to adhere to this, however, an amendment of the Comitology decision in this sense seems necessary. This requirement fits in with the General Framework by serving the objectives of coherence, transparency, and especially accountability.

[15]See Bergström 2005; Christiansen and Larsson 2007; Joerges & Vos 1999; van Schendelen 1998.

6.6 The General Framework and General Principles of European Law

It is important that any decision adopted on the basis of the General Framework should comply with general principles of Community law, in particular the precautionary principle, the proportionality principle, and the subsidiarity principle.

6.6.1 *Precautionary Principle*

Today the precautionary principle is an important pillar of food safety regulation. The application of this principle is a source of much debate and controversy in Europe, and its application leaves much to be desired in terms of consistency and clarity.[16] The GFL labels the precautionary principle as a general principle of food safety and is defined in Art. 7. Notwithstanding this definition, there is still much unclarity about the precise significance of the precautionary principle. In its landmark case *National Farmers' Union* the European Court of Justice (ECJ) gave a broad definition to the precautionary principle stating that

> [w]here there is uncertainty as to the existence or extent of risks to human health, the institutions may take protective measures without having to wait until the reality and seriousness of those risks become fully apparent.[17]

Whilst other studies have already discussed the precautionary principle in much detail (Renn et al. 2003), this book aims to give the precautionary principle a place in the process of food safety governance, recognising that the principle needs to be applied throughout the whole process. Nevertheless, in its Communication on the Precautionary Principle of 2000, the Commission emphasised its view that the precautionary principle should be regarded as a risk management principle (CEC 2000a). It argued that

> the precautionary principle is particularly relevant to the management of risk. The principle, which is essentially used by decision-makers in the management of risks should not be confused with the element of caution that scientists apply in their assessment of scientific data (CEC 2000a: summary, para 4).

Also the ECJ seems to see the principle foremost as a principle of risk management, although phrased in more flexible wording and referring to it as being

[16] E.g. de Sadeleer (2006); Corcelle (2001); Marchant and Mosman (2004); Forrester and Hanekamp (2006); Alemanno (2001); Douma (2002); see also: de Sadeleer (2001a), (2001b); Scott (2004); Ladeur (2003); Faure and Vos (2003).

[17] E.g. Case C-157/96, *The Queen v Ministry of Agriculture, Fisheries and Food* [1998] ECR I-02211, para.63.

an integral part of the decision-making processes leading the adoption of any measure for the protection of human health.[18]

Yet, recent thinking in legal circles point out that from a legal point of view, nothing precludes that the risk assessment stage has to be carried out in accordance with the obligations stemming from the precautionary principle. We thus argue that in order to deal effectively with uncertainty, ambiguity, and ignorance, assessors should apply precaution at an early stage (de Sadeleer 2006: 148).

6.6.2 Proportionality Principle

The proportionality principle says that

> any action of the Community shall not go beyond what is necessary to achieve the objectives of the Treaty (Article 5 (3) EC Treaty).[19]

In this way, it has particular relevance for risk governance measures, protecting human health. The proportionality principle has been developed in the case law of the European Court of Justice in the context of trade hindering measures adopted by the Member States. In particular the ECJ developed a threefold test to examine the validity of the measures adopted by the Member States and, in a later stage, the measures adopted by the Community institutions. The proportionality of the measures is thus judged by looking at the aim and nature of the measure. Questions to examine include whether or not:

(1) A measure is necessary in order to protect one of the recognised interests (such as protection of health and the environment),
(2) The measure is the least restrictive of trade, and
(3) The imposed restrictions are proportionate to the aim pursued.[20]

Examination of the early case law of the Court of Justice revealed that in the field of free movement of goods the proportionality principle, as developed by the Court, had already included a kind of precautionary principle long before the precautionary principle appeared in the Community context as a "true" principle (Scott & Vos 2002: 25). It can thus be said that the precautionary principle "grew out" of the proportionality principle, before it was finally recognised by the ECJ as an autonomous principle applying also to health issues.[21]

---

[18] Case C-236/01, *Monsanto Agricoltura Italia* ECR [2003] I-8105, paragraph 133.

[19] See e.g. Protocol on the application of the principles of subsidiarity and proportionality.

[20] See de Sadeleer (2006, 148): referring to van Zwanenberg and Stirling (2003: 49).

[21] E.g. Case T-70/99, *Alpharma v Council* [2002] ECR II-3495 and T-13/99 *Pfizer Animal Health v Council* (Case T-13/99 *Pfizer Animal Health v Council* [2002] ECR II-3305 as well as joint Cases T-74/00 *Artegodan v Commission* (Joined Cases T-74/00, T-76/00, T-83/00, T-84/00, T- 85/00, T-132/00, T-137/00 and T-141/00, *Artegodan GMbH and Others v Commission* [2002] II-ECR 4945).

6.6.3 Subsidiarity

The principle of subsidiarity is a very much debated principle, too. It is laid down in the EC Treaty in Article 5 (2) which states:

> In areas which do not fall within its exclusive competence, the Community shall take action, in accordance with the principle of subsidiarity, only if and in so far as the objectives of the proposed action cannot be sufficiently achieved by the Member States and can therefore, by reason of the scale or effects of the proposed action, be better achieved by the Community.

The importance of this principle for the general risk governance is clear: the Community institutions should not exercise their powers in a way considered detrimental to the Member States. Yet, it will be clear that in view of the objective of free movement of foods, it will be likely that the Community will legitimately exercise its powers ensuring free circulation of those goods in the whole Community market. As we know, the subsidiarity principle clearly dictates that Member States should not be excluded from the process of creating a European Union based upon the rule of law, democratic principles, and solidarity. In this manner, one could say that observing carefully the procedural element of decision-making taking into consideration the level at which decisions are taken, how and in what way they are drafted, is also a means of implementing the philosophy of subsidiarity (Dehousse 1994: 124 pp.). Therefore mechanisms which provide for co-operation between all the levels concerned might address Member States' concerns for unnecessary Community activities and hence respect the subsidiarity principle (Vos 1999). Where the Interface Advisory Committee proposed by the General Framework provides for the possibility to also include Member States as ad hoc members, this can be regarded as implementing the subsidiarity principle. The same applies to the opportunity for Member States to express their views through the Internet Forum.

6.6.4 Good Governance

The General Framework directly addresses the five principles of good governance identified in the European Commission's 2001 White Paper on European Governance. With regard to the principle of *openness*, the Paper prescribes that EC institutions "should work in a more open manner" and "actively communicate about what the EU does and the decisions it takes" (CEC 2001a: 10). As has been made clear in the course of this chapter, one of the primary objectives of the innovations proposed by the General Framework is to increase the transparency of food safety governance especially during the crucial steps at the interface of scientific assessment and political decision-making, requiring that all relevant interface communications should be made accessible to interested parties and the wider public through the Internet Forum. This applies equally to the principle of *participation*, which is addressed as a major objective of all steps of food safety governance outlined in the General Framework, and specifically supported through the creation of the Internet Forum and an Interface Committee that involves stakeholder representatives (see Chap. 7 for a more detailed discussion). By rendering the interaction of assessment and management less opaque and more open to critical observation and debate, the General Framework also helps to realise the principle of *accountability*, requiring that "each of the EU institutions must explain and take responsibility for what it does in Europe" (CEC 2001a: 10). Through the recommendation to re-consider decision-making practices at the stage of the comitology procedure, the General Framework also follows the objective of increasing the clarity and responsibility of decisions made by the Member States, required by the accountability principle. Furthermore, the General Framework takes into account the principles of *effectiveness* and *coherence* by proposing a more effective and appropriate distinction of threats through screening, and by establishing an interface structure to render the co-ordination between assessment and management more systematic and effective.

6.6.5 Good Administration

Finally, the General Framework builds on the principle of good administration as one of the basic rights of citizens protected by European law. The principle is based on Article 41, on the right to good administration of the Charter of Fundamental Rights of the European Union, which was formally proclaimed by the Heads of State and Government at the Nice European Council and now referred to as legally binding by the (yet not ratified) Lisbon Treaty.[22] Although still formally non-binding at the present state, the Charter of Fundamental Rights may have visible effects, as its provisions can be used by national and European courts to interpret national and Community legislation in conformity with the Charter, especially with regard to provisions directly concerning the behaviour of public authorities such as Article 41 (cp. van Gerven 2005: 125). Furthermore, the right to a good administration may be called upon by citizens by referring cases of maladministration in the activity of Community institutions or bodies, to the European Ombudsman, a right set out by Article 43 of the Charter of Fundamental Rights. The right to good administration is also remarkable through the fact that it applies not only to EU citizens, but also to every person coming into contact with the Union's institutions and bodies. The principle of good administration is established as guidance to the administrative behaviour of Community institutions and bodies, demanding their relations with the public. The first two paragraphs of Article 41 set out the content of the principle of good administration, establishing the principles of impartiality, fairness, and reasonable time limits, giving every person the right to be heard prior to any measure which might affect him or her adversely, and establishing the obligation on Community

[22] Article 6 of the Treaty on European Union as amended by the Treaty of Lisbon (OJ 2008, C115/13).

institutions and bodies to give reasons for their decisions (also enshrined in Article 253 of the Treaty.[23]) The third and fourth paragraphs of the Article concern the compensation of damage caused by the EU institutions and the right to make written inquiries and receive answers in any of the languages of the Treaties.

Departing from this legal principle, various attempts have been made to give substance to the exact meaning and application of the right to good administration. In this vein, the contents of this Article have been spelled out in a European Code of Good Administrative Behaviour, drafted by the European Ombudsman and approved by means of a resolution of the European Parliament on 6 September 2001. The code details the rules of good administrative behaviour that EU institutions and bodies, and their administrations and officials should respect and abide by. Apart from imposing general principles of lawfulness, proportionality, objectivity, fairness, impartiality, and absence of discrimination and abuse of power, the Code prescribes in its Article 16 the right to be heard and make statements. The Article prescribes that in cases where the rights or interests of individuals are involved, officials of European Institutions shall ensure that the rights of defence are respected, allowing every member of the public the right to submit written comments in cases where a decision affecting his or her rights or interests are affected. Furthermore, Article 18 on the duty to state the grounds of decisions, places the European institutions under the obligation to state the grounds of decisions that may adversely affect the rights or interests of a person, indicating the relevant facts and the legal basis of a decision (see also Article 253 EC). In addition, officials of European institutions are obliged to provide citizens who expressly request it with an individual reasoning for decisions.

In addition to the general rights and obligations, the European Commission has also specified its own rules of good administrative behaviour in its relations with the public (adopted on 13 September 2000), following initiatives to improve its administrative practices triggered by its White Paper on Administrative Reform, adopted after the resignation of the Santer Commission in 1999. Adding to the general obligations of European institutions and their officials outlined above, the Code prescribes that in cases where Community law provides that interested parties should be heard, Commission staff shall ensure that an opportunity is given to them to make their views known. Furthermore, the obligation is established that a Commission decision should clearly state the reasons on which it is based, requiring full justification for decisions as a general rule. Moreover, Article 3 of the Code sets out that any interested party who expressly requests a detailed justification shall be provided with it.

Many of these obligations of "good" or "sound" administration have been developed by the ECJ. Important for the General Framework is, in particular, the duty

[23] Article 253 EC sets out that "Regulations, directives, and decisions adopted jointly by the European Parliament and the Council, and such acts adopted by the Council or the Commission, shall state the reasons on which they are based and shall refer to any proposals or opinions which were required to be obtained pursuant to this Treaty".

for the Community institutions "to examine, carefully and impartially, all the relevant aspects of the individual case".[24] This means for example that the scientific assessment must be made on the basis of scientific advice founded on the principles of excellence, transparency and independence in order to guarantee the scientific objectivity of the measures adopted and to preclude any arbitrary measures.[25]

The significance of the principle of good administrative behaviour and the codes of conduct presented above for the General Framework is twofold. Firstly, by establishing the right to persons to be heard and the obligation of Community institutions to state reasons for their decisions, thus providing a guidance for the interaction between the expression of views in the Internet Forum and reactions by the Community institutions, in particular the Commission and EFSA. As mentioned above, these provisions establish the obligation to take into account the views expressed in the Internet Forum and to give reasons for decisions in relation to these views, especially in cases where the interests of individual persons are obviously affected. Whereas there would be no formal reporting mechanism from the Internet Forum to EFSA or the Commission, and the Internet Forum would not be able to directly determine the agenda of the Interface Committee, an obligation is established to take into account and discuss interests and concerns expressed by stakeholder groups and individual citizens through the online function. Secondly, the proposed institutional innovations can also be seen to further implement the objectives established by the right to good administration and the codes of good administrative behaviour, by giving both civil society actors and individual citizens an accessible instrument to make their views known, and to provide a forum for the Commission and EFSA to state the reasons behind their decisions through the increased transparency of communications at the interface between assessment and management.

6.6.6 The Principle of Non-Delegation of Powers (Meroni Doctrine)

In this context, specific attention is also given to the principle of non-delegation, as expressed in the so-called "Meroni" doctrine. The doctrine is still *the* dominant argumentation framework both in legal and political debates for restricting tendencies of functional decentralisation in the institutional structure of the EC to the degree of giving only very specific and limited powers to independent agencies (such as EFSA) and other bodies that are independent of the Commission.

[24] Case T-13/99, para. 171, with reference to Case C-269/90, *Technische Universität München v. Hauptzollamt München-Mitte* [1999] ECR I-05469, para. 14.

[25] Ibid.

This doctrine was inspired by the case law of the European Court of Justice of the late 1950s.[26] In the *Meroni* cases, the Court rejected the transfer of sovereign powers to subordinate authorities outside the EC institutions and ruled that only "clearly defined executive powers" could be delegated, the exercise of which was to remain at all times subject to Commission supervision. Although the *Meroni* judgments related to the Treaty establishing the European Coal and Steel Community (ECSC), their applicability to the EC Treaty has been generally accepted (see Lenaerts 1993). This case law would suggest that only "strictly executive powers" may be delegated to bodies other than the European institutions as only then the institutional structure of the Community would remain intact.[27] Although over the years some pro-delegation voices have been heard in the Commission, it is currently still the prevailing opinion, known as the "Meroni" or "anti-delegation" doctrine that no discretionary powers can be delegated to committees or agencies that are created within the Community's institutional structure (see Majone 2002: 330–331). This is also the reason for the "intermediate option" being the General Framework's preferred option (see Sect. 6.4.2), with the advisory nature of the Interface Advisory Committee fully respecting this *Meroni* or anti-delegation doctrine. The *Meroni* doctrine is relevant for the application of the General Framework with regard to the following aspects:

- *Terms of reference/evaluation*. The doctrine may have implications for the step of setting the terms of reference, as the intention is to transform the specification of these terms from a closed process within the Commission into a co-operative exercise that is shared with assessors and stakeholders, and which may be transferred to an external forum composed of these three actor groups. This proposed change is not seen as infringing on the doctrine, as the setting of the terms of reference does not predetermine the outcome of assessment, and even less of the decision taken later on at the step of management. Nevertheless, if it should be felt that the *Meroni* doctrine interferes with the setup of a new organ deciding on the terms of reference, as this takes away relevant functions of risk analysis from the Commission, the "intermediate" and "minimal" options take account of such concerns. It is, therefore, also up to the interpretation of the *Meroni* doctrine

[26] Case 9/56, *Meroni & Co., Industrie Metallurgische S.p.A. v High Authority* [1957–1958] ECR 133 and Case 10/56, *Meroni & Co., Industrie Metallurgische S.p.A. v High Authority* [1957–1958] ECR 157.

[27] This case law would suggest that the following conditions apply to the admissibility of transferring sovereign powers to subordinate authorities outside the EC institutions: the Commission cannot delegate broader powers than it enjoys itself; only strictly executive powers may be delegated; discretionary powers may not be delegated; the exercise of delegated powers cannot be exempted from the conditions to which they would have been subject, had they been directly exercised by the Commission, in particular the obligation to state reasons for decisions taken, and judicial control of decisions; the powers delegated remain subject to conditions determined by the Commission and subject to its continuing supervision; and the institutional balance between the EC institutions must not be distorted; see Vos (2003).

which of the options to chose for this step. Equally, the doctrine needs to be considered in relation to the step of evaluation, which is also recommended as a task to be undertaken in cooperation between managers, assessors and stakeholders, in the Interface Committee that is independent from the Commission. However, the General Framework takes account of this concern in defining the task of the Interface Committee as a *purely advisory* one, which does not interfere with the full responsibility of the Commission for the decision about the outcomes of evaluation and the eventual conduct of management.

- *Assessment.* The doctrine clearly has strong implications for the conduct of assessment, as tasks within this stage of food safety governance can only be allocated to EFSA as far as they fall within the sphere of risk assessment as defined by the General Food Law, and can thus be separated from functions of risk management falling under the responsibility of the Commission. This requires clarifications in some cases such as the presumption of prevention (in which the application of crisis management mechanisms is understood as a function of management), precautionary assessment (which is understood not to interfere with the final responsibility of the managers to apply the precautionary principle), and concern assessment (which refers to the gathering of information about socio-economic concerns, but not to their evaluation). Furthermore, it is understood that the choice of one of the approaches to assessment for a particular case of food safety governance does not preclude the choice of a particular management strategy and does therefore not interfere with the autonomous decision of the managers of selecting, ranking, choosing, and implementing particular options to deal with a given food safety threat.
- *Management.* This step does not pose a particular problem in the light of the non-delegation doctrine, as decision-making is fully left as a responsibility assigned to the Commission (and the Member States), as set out in the General Food Law.

6.7 The General Framework and WTO Law

As stated in the introductory remarks, one of the primary objectives of the General Framework is to achieve a full compatibility of food safety governance procedures with requirements at international level, especially in the framework of WTO agreements. The General Framework now, we argue (and, in particular, the way it proposes to carry out the stage of *assessment*) might be interesting for the EC as it offers a potential manner to make those decisions which are adopted according to the General Framework, "WTO compatible". In this context, the added value of the General Framework is demonstrated through its *objectivation or rationalisation of non-scientific values*. As has been shown in previous chapters, these are subjected to a rigorous test through scientific principles, first at the stage of screening, and then addressed through setting up terms of reference and assessment, taking into account, in a systematic manner, the sources of scientific uncertainty and socio-political ambiguity recognised by assessors, managers, and stakeholders. Seen in this light, the concern assessment proposed in the General Framework, which is about gathering evidence about concerns through scientific methods (and hence not about expressing opinions on that evidence), could fall under the concept of "scientific evidence" as interpreted by the WTO Appelate Body in Japan Apples, where it found scientific evidence to be: "evidence gathered through scientific methods, excluding by the same token information not acquired through a scientific method".[28] In this manner, measures based on concern assessment and drawing on non-scientific values could potentially be regarded as science-based in the WTO context.[29]

6.8 Conclusions

The General Framework proposes a limited set of optional institutional innovations referring mainly to the improvement of capacities of EFSA and the better co-ordination of management and assessment. These proposals are in line with insights gained from empirical research and documents from both EFSA and the Commission calling for improved communication and transparency at this interface. In this sense, three options for the design of interface organs have been presented, one of which – the "intermediate" option combining the Internet Forum with an Interface Advisory Committee – is proposed as the preferred option. As outlined above, these options have grown out of discussions with risk managers, risk assessors, and representatives of both industry and NGOs in a series of workshops in which initial proposals were presented and afterwards revised in the light of the comments and suggestions received (see Chap. 11 for details of this workshop-based review and feedback process). These proposals are therefore not just the product of a purely academic exercise, but reflect and integrate the viewpoints of policy practitioners from both European and Member-State levels.

It was stated from the outset that one of the major aims of the General Framework is to be fully compatible with the existing institutional structures of EU food safety regulation, the general principles of European law, and the requirements established through case law of the European Court of Justice and international agreements especially in the framework of the WTO. This chapter showed that the General Framework is firmly based on the objectives of improving the application of the principles of good governance and good administration, while being fully compatible with the principles of subsidiarity and proportionality. Furthermore, it is important to stress that the General Framework proposals not only are compatible with international requirements, but they can be used to establish more solidly the

[28] *Japan Apples* (Panel) para. 8.92.

[29] See for an excellent analysis of the SPS agreement, Scott (2007); see about WTO law in general, van den Bossche (2008).

compliance of food safety governance decisions with requirements established through the SPS agreement. Finally, as stated in previous chapters, the General Framework contributes to the understanding of scientific uncertainty and the clarification of the suitability and necessity of measures (to be) taken, thus constituting a major step forward in the consistent and transparent application of the precautionary principle in European food safety governance.

Chapter 7
A Structured Approach to Participation

M. Dreyer and O. Renn

7.1 Introduction

All previous chapters have already touched on the topic of participation. Chapter 6 has pointed out the core of the participatory design of the food safety governance framework this book proposes: It consists of food safety interface institutions – the Interface Committee and the Internet Forum – which are destined to function as intermediaries between science, policy, and civil society. The present chapter will provide a condensed presentation of the envisioned participatory design of the governance of food safety and, in doing so extend considerations on how to tailor participation to the purposes served at the different governance stages. Firstly, it will highlight the special value that is assigned to the *interface institutions* as formal mechanisms for putting the idea of inclusive governance advocated by this book into practice. Secondly, this chapter will present a *guiding tool* designed to assist the Interface Committee, or the European Commission solely (if no Interface Committee were to be set up), to specify whether it is required to resort to more extensive participation in a given case, i.e. to select *additional* participatory processes (extending beyond the inclusion of stakeholders and the wider public through the Interface Committee and web-based consultations and deliberations). This guiding tool, which will be set out in more detail below, distinguishes between different *purposes* of participation, specific to the respective governance stage, and different levels of *intensity* of participation depending on the levels of uncertainty and ambiguity.

7.2 Participation Through Food Safety Interface Institutions

This book recognises the idea of *inclusive governance* as a necessary (although not sufficient) prerequisite for tackling food safety problems in both a sustainable and acceptable manner and, consequently, imposes an obligation to ensure the early and meaningful involvement of a diversity of social groups (Jasanoff 1993). Inclusive governance is based on the assumption that affected and interested parties have

M. Dreyer and O. Renn (eds.), *Food Safety Governance*,

DOI: 10.1007/978-3-540-69309-3_8, © Springer-Verlag Berlin Heidelberg 2009

something to contribute to the process of food safety governance, and that mutual communication and exchange of ideas, assessments, and evaluations improve the final decisions rather than impede the decision-making process or compromise the quality of scientific input and the legitimacy of legal requirements.[1] As the term 'governance' implies, analysing and managing food safety threats cannot be confined to private companies and regulatory agencies. Rather it involves a wider array of actors: political decision-makers, scientists, economic actors, and civil society actors.

As set out in detail in Chap. 6, the General Framework advocates the setting up of 'food safety interface institutions' in order to improve the co-ordination between these key actors in the governance of food safety. These interface institutions present platforms for deliberation on major elements of the governance process. The *Internet Forum* is the most inclusive of the proposed interface institutions as it offers a deliberation platform with open public access (however, in order to keep the appraisal of the Forum's discussions practicable, detailed posting should be conceded only to accredited stakeholders). Not only corporate and civil society actors, but also those responsible for management at Member State level, and scientific experts affiliated to the national Competent Authorities could use this deliberation forum to engage with the diversity of subjects. The Internet Forum is inclusive also in that it provides the opportunity to deliberate on all of the major elements underlying governance outcomes including the referral details, the screening results, the terms of reference, the assessment results, and the evaluation conclusions (Chap. 6 provides detailed discussion thereof).

To create transparency on these elements means to subject the *reasons* of decision-making on food safety problems to public scrutiny. By inviting and expecting participants to not merely state their opinions but to also exchange views, i.e. to discuss each others' standpoints and arguments, the Internet Forum extends beyond a mere consultation process: it is designed to provide the Commission and the proposed Interface Committee not only with *individual* feedback but also with feedback based (at least in part) on discussion, reflection, and persuasion, i.e. with opinions mutually informed by a diversity of views. Hence, the Internet Forum ties in with the increasing use of the Internet for documentation and consultation by both EFSA and the Commission[2], but is aimed at providing, in addition, a forum for deliberation. Certainly, the breadth and intensity with which individual cases would be discussed through the Internet Forum can be expected to vary greatly, very much depending on the potential for conflict that might be implied in the cases. In that sense, the Internet Forum could act as both an *entry point* at major governance stages of a diversity of viewpoints into the governance process, and a *signal* for highly controversial issues with a great potential for social mobilisation.

The recommendation of this book is to complement the Internet Forum by a food safety interface institution which brings managers, assessors, and key stakeholders together in a *committee structure* at two stages in the governance process:

---

[1]See similar arguments in Webler (1999) and Renn (2004).

[2]For an overview of the recent developments in stakeholder involvement in EU food safety governance, see Wendler and Vos (2008).

at the framing stage and the evaluation stage. It was underlined in earlier chapters that the interlinkages between the scientific and political aspects of food safety governance are particularly strong when questions and tasks are defined in relation to a given food safety threat (i.e. when the problem is framed) and when tolerability and acceptability judgements are made (i.e. when the problem is evaluated). This 'hybrid' character of framing and evaluation is likely to explain, at least in part, the need for improved interaction between assessors and managers in the performance of these activities, which was expressed by several EU-level and Member State assessors and managers whose views were elicited in the study of the governance systems at EU-level, and in France and Germany where assessment and management responsibilities are allocated to different institutions (Dreyer, Renn, Borkhart, & Ortleb 2006, cp. Sect. 1.2.1 this volume). Judgments on facts and values are of equal importance in framing and evaluation. We recommend taking this fact into account by institutionalising a direct face-to-face exchange between the Commission, EFSA, and selected stakeholders about setting the terms of reference and evaluation. The deliberations of the Interface Committee (in one of its two variants) would draw upon stakeholder perspectives sought through the Internet Forum in order to take account of a broader range of viewpoints (unlike the 'Steering Committee' the 'Advisory Committee' would not be convened for every case but only for cases identified as specifically challenging and it would act merely in an advisory function; see Chap. 6 for a detailed account of the way in which the two committee options differ).

7.3 A Guiding Tool For Deciding on Extended Participation

As has already been mentioned above, specific cases might require that participation through the interface institution(s) be complemented by additional participatory processes. The proposed governance framework envisions a *proceduralisation* of decision-making over any possible extension of the scope of participation and about the selection of appropriate processes: If an Interface Committee is set up, it is part of the mandate of this body to advise on this matter at the stages of framing and evaluation in consideration of the specific case and the given context and the overall socio-political climate. In all cases it will be the responsibility of the Commission to take the decision over the necessity for additional participatory processes.

Aspects that could inform this decision-making process might possibly be derived from the Internet Forum and from the stakeholders who sit on the Interface Committee and can act as *sensitivity sensors* for highly controversial issues which call for broader participation. In addition, those consultative stakeholder bodies which have been established in recent years might be of some assistance in this respect, i.e. EFSA's Stakeholder Consultative Platform which had its inaugural meeting in October 2005, the European Commission's Advisory Group on the Food Chain and Animal and Plant Health established in August 2004, and DG SANCO's Stakeholder Dialogue Group, which was created in December 2007.

The primary task of these bodies is, however, to consult on broader policy and strategic issues and on questions with a more general relevance for risk assessment. Hence, they will deal with individual food safety problems only in exceptional cases. Yet in these exceptional cases, their discussions appear to deal also with particular aspects of framing and evaluation (Wendler & Vos 2008). It would, therefore, be important for the discussion results to be taken into consideration in the Interface Committee deliberations.

While these sources of information already have a great potential for facilitating decision-making around the need for broader participation, the General Framework, in addition, offers a *default assumption* for decision guidance: It presupposes that a higher degree of participation will also be required under the conditions of *high levels* of scientific uncertainty and socio-political ambiguity. This corresponds with the central institutional idea that the Interface Advisory Committee is not convened for every case at the stages of framing and evaluation (in contrast to the Interface Steering Committee) but only for specifically challenging cases, including those cases where screening has identified the conditions of scientific uncertainty and/or socio-political ambiguity.

In short, the guiding tool it offers for deciding on more extensive participation distinguishes between *different levels of intensity* and also *diverse purposes* of participation (illustrated in a schematic form in Table 7.1 below). Intensity is linked to the likelihood of major societal debate or conflict surrounding the threat under review which is assumed to be higher under the circumstances of high levels of scientific uncertainty and socio-political ambiguity (on which the screening stage

Table 7.1 A structured approach to participation

Governance stage	Style of discourse	Purpose: as a contribution to	Institutionalised participation	Additional participatory processes
Framing	Design	Drawing up the terms of reference	Via the *Internet Forum* throughout the governance cycle	*Procedurally*, context-dependent, and specified at the stages of framing and evaluation
Assessment	Epistemic	Gathering knowledge and information		
Evaluation	Reflective	Value-based judgements on tolerability or acceptability	At the stages of framing and evaluation: via stakeholder representation on the *Interface Committee*	*Prima facie default*: high levels of scientific uncertainty and/or socio-political ambiguity require extended participation
Management	Practical	Selection of appropriate measures		

provides preliminary information, see Sect. 4.2). The different purposes of participation are being served at the different stages in the governance process and must be taken into account in the selection of appropriate participation processes.

The question of what follows the requirements for extended participation will be discussed with regard to each of the four major governance stages. The purpose of participation will be discussed in terms of the type of discourse which is identified as being generic to each respective stage.[3]

7.3.1 Participation During Framing

The type of discourse that is generic to the framing stage is called *design discourse*. This discourse (involving the Interface Committee, if set up) is aimed at setting the terms of reference including the scope, focus and design of assessment and at specifying the way (breadth, concrete procedures) in which stakeholders and/or the wider range of public are included in the assessment process beyond the formalised engagement mechanisms (i.e. the Internet Forum). Only in those cases where screening identifies high degrees of scientific uncertainty and/or socio-political ambiguity would it be advisable to complement stakeholder participation through the Internet Forum (where the referral details and the screening results are documented) and the Interface Committee (if set up) by additional participatory processes. Appropriate procedures that could be used in a design discourse include formal hearings of relevant commercial and civil society groups (see Annex 1B for a short portrayal of this participatory instrument), open space conferences, and public forums.

7.3.2 Participation During Assessment

The type of discourse that is generic to the assessment phase is entitled *epistemic discourse*. It comprises communication processes, where experts of knowledge (not necessarily scientists) grapple with the clarification of a factual issue (see Annex 1A for a short portrayal of some participatory instruments particularly suited for an epistemic discourse). The goal of such a discourse is the representation and explanation of a phenomenon as close to reality as possible. By knowledge we refer to *systematic* knowledge collected by established means of natural and social sciences and *experiential* knowledge collected by interactive techniques such as hearings or focus groups. Both types of knowledge are important for describing what we generally know about the threat (or about a set of functional equivalents to a threat source) and what we have learned in dealing with the threat or a similar threat source in the past.

[3]The labels for these different discourse types were first introduced by Renn (1999).

Subject to the provisions of framing, civil society actors and also the wider public may contribute to broadening and refining the infrastructure of knowledge and information, upon which evaluation and management decisions draw, also beyond the Internet Forum (where the terms of reference would be documented) through face-to-face methods of consultation and/or deliberation-based interactive elicitation. The conditions of high levels of scientific uncertainty and socio-political ambiguity in the first place would suggest such extended participation:

- When a given threat is approached by a precautionary assessment, stakeholders should be asked to administer their specific knowledge regarding the likely consequences of the product/process/practice in question that carries a certain threat. The more uncertain the given threat is, the more a communicative exchange among experts of a great diversity of disciplines and also practical backgrounds is required to reach a coherent description and explanation of the phenomenon. Frequently, these discourses can only show the range of the methodically still justifiable knowledge, i.e. define the boundaries between the absurd and the possible, between the possible and the likely, and between the likely and the certain. Methods for this type of involvement include the Delphi and Group Delphi method, scientific consensus conferences and meta-workshops (Turoff 1970; Webler, Levine, Rakel, & Renn 1991). Under conditions of high scientific uncertainty, stakeholders should also be invited to engage in a comparative review and administer their specific knowledge in relation to a range of *alternative options* (i.e. functional equivalents) to the product/process/practice in consideration. The realm of knowledge needed to characterise uncertain threats expands the scope of traditional risk analysis and includes expertise about social benefits associated with the threat or its alternatives, about possible substitution pathways, potential for using 'forgiving' technologies, etc. Methods such as stakeholder surveys, qualitative interviews, focus groups, and public hearings are most appropriate for this task.
- When a given threat is approached by a concern assessment, engagement with stakeholders is vital to elicit, as widely as possible, the *concerns, perspectives, and preferred options* that the relevant social groups, on the basis of their specific knowledge and information, have regarding the case under review. If the assessment drawing on the contributions and deliberations in the Internet Forum reveals that there is much debate, even in the wider public, and a high potential for social conflict involved, it might be necessary to also conduct face-to-face inquiries among different groups and representatives of the wider public. Methods for this type of involvement include focus groups, stakeholder interviews, hearings and other interactive elicitation methods such as value tree analysis, option mapping, and others.

It is important to note, that it is *not* the task of stakeholders and representatives of the wider public at the assessment stage to deal with normative questions pertaining to the tolerability or acceptability of either the threat itself, different strategic options (a set of products/processes/practices which are possible alternatives to the option in question), or management measures for tackling the threat. These normative issues are part of the evaluation and management phases. They are based on value judgements about what is 'desirable' rather than what is 'true'.

7.3.3 Participation During Evaluation

The type of discourse that is generic to the evaluation phase is named *reflective discourse* (see Annex 1B for a short portrayal of some participatory instruments which could be used for a reflective – or practical – discourse). This discourse comprises communication processes dealing with the interpretation of factual issues, the clarification of preferences and values, and a normative judgement of tolerability or acceptability. Reflective discourses are mainly suitable for balancing pros and cons, weighing the arguments and reaching a balanced decision on the basis of the epistemological discourse and social values and preferences.

The purpose of stakeholder engagement here is to ensure that all values and preferences are included in the weighing procedure, and that the final judgement reflects the societal balance between innovativeness and caution. The stakeholders sitting on the Interface Committee (if set up) would re-convene with the managers and assessors during this phase and use the new knowledge from the assessment to draw normative conclusions about the threat in consideration. Part of the evaluation process would be to draw on the Internet Forum (where the assessment results and the (draft) evaluation conclusions would be documented) to judge the need for more comprehensive engagement involving additional stakeholders and/or the wider public. Again, the conditions of high levels of scientific uncertainty and socio-political ambiguity in the first place would suggest a more elaborate participation programme:

– When scientific uncertainty is implied with a given threat, the central question is: How can one judge the severity of a situation when the potential damage and its likelihood are unknown or highly uncertain? In this dilemma, the Interface Committee or the Commission on its own (if the Interface Committee were not to be set up) may have to include all of the relevant stakeholders in a face-to-face participatory deliberation and ask them to find a consensus on the extra margin of safety (or alternative measure) in which they would be willing to invest in exchange for avoiding potentially catastrophic consequences. This type of deliberation relies on a collective reflection about balancing possible over- or under-protection. If too much protection is sought, innovations may be prevented or stalled; in case of too little protection, society may experience unpleasant surprises. The classic question of 'how safe is safe enough' is replaced by the question of 'how much uncertainty and ignorance are the main actors willing to accept in exchange for some given benefit'. It is recommended that policy makers, scientists, and representatives of all relevant social groups (including the Interface Committee if set up) take part in this type of extended face-to-face discourse.

It is also essential that the discourse should not just be preoccupied with the threat under review but that it also considers potential alternatives, social benefits, sustainable practices, and other related aspects. Methods for this type of extended involvement include round tables, open space forums, negotiated rule-making exercises, or mediation.[4]

– Threats characterised by high ambiguities require the most inclusive strategy for participation since not only directly affected groups have something to contribute to this debate, but also those indirectly affected. Resolving ambiguities in food safety debates necessitates a platform where competing arguments, beliefs and values are openly discussed. The opportunity for resolving these conflicting expectations lies in the process of identifying common values, defining different angles or perspectives allowing people to apply their own vision of a 'good life' to judging the tolerability or acceptability of threats, without compromising the vision of others. Under the condition of high levels of socio-political ambiguity and a great potential for social conflict and mobilisation it is recommended complementing the deliberation through the Interface Committee (if set up) and the Internet Forum by face-to-face participatory deliberation with citizen involvement. Available sets of deliberative processes in which a randomised or deliberately stratified group of citizens work to scope and explore the issues and options in contention include citizen panels, citizen juries, consensus conferences, ombudspersons, citizen advisory committees, and others.[5] In addition, classic stakeholder engagement processes such as hearings might accompany the public participation program.

7.3.4 Participation During Management

The type of discourse that is generic to the management phase is called *practical discourse* (see Annex 1B for a short portrayal of participatory instruments which could be used for a practical – or reflective – discourse). It comprises communication processes aimed at the identification, assessment, and selection of different management measures for reducing and managing 'intolerable threats' or 'tolerable but not acceptable' threats. The term 'practical' refers to the nature of decision-making, i.e. the different steps outlined in Sect. 5.3. The practical discourse looks at the variety of possible interventions, addresses the pros and cons for each measure or package of measures, and suggests a set of measures that appear to be effective, efficient, and fair. The main purpose of participation is here to ensure that relevant

[4]For a discussion of the use of these methods in the environmental field see: Susskind, Richardson, and Hildebrand (1978); Amy (1983); Moore (1996); Owen (2001); Gregory, McDaniels, and Fields (2001).

[5]See Kasemir et al. (2003); Joss (1999); Armour (1995); Renn, Webler, Rakel, Dienel, and Johnson (1993); Kathlene and Martin (1991); Dienel (1989); Crosby, Kelly, and Schaefer (1986); for reviews see: Hagendijk and Irwin (2006); Rowe and Frewer (2000); Lynn (1990).

knowledge and different preferences are being considered in the conclusions on the selection of one or more management measures. If set up, the Interface Committee would give advise on participation procedures in this discourse with the Commission taking the decision. It is recommended that participatory deliberation reaching beyond the Internet Forum (where the evaluation outcome including the most appropriate management approach would be documented) should be employed at the stage of management when a high level of scientific uncertainty surrounds a given case, and/or under the condition that not only is the threat itself contested but socio-political ambiguities extend to the selection of management measures.

- For highly uncertain threats it is advisable to have stakeholders involved in an exercise to balance pros and cons associated with each of the potential measures. Measures that increase resilience or robustness (as advocated by a precautionary approach) are often inferior to cost-minimisation strategies when cost–benefit analysis or other formal balancing techniques are applied. Therefore the question of what methods to use when balancing pros and cons for evaluating a variety of measures should be a major topic of the stakeholder discussions. It is recommended that policy makers, representatives of major stakeholder groups, and experts on the impacts of each measure should take part in the discourse. Methods for this purpose include negotiated rule-making exercises, mediation, or mixed advisory committees including scientists and stakeholders.[6]
- High socio-political ambiguity may lead to very different visions between social groups of how to address these ambiguities in form of management measures. If the measures are also highly contested, it seems advisable to organise a broad societal discourse about the appropriateness of these measures and the best way to achieve a consensus or an agreement on the measures to be taken. However such a discourse is conducted, the design of the participatory procedure should allow for a high degree of *representativeness* on the part of participants in relation to interested and affected parties in the wider society. The methods for addressing ambiguity in the evaluation process are also appropriate for handling ambiguity in the selection of management measures and hence include citizen forums, citizen panels, citizen juries, consensus conferences, ombudspersons, citizen advisory commissions, and similar participatory instruments in addition to classic stakeholder engagement processes.

All four forms of discourse require the design of the participatory procedures to display these basic features:

- A good level of *transparency* from the point of view of third parties, in documenting how specific inputs relate to the decision on one or more management measures
- No *constraints* as to the way in which participants may express themselves

[6]See Stolwijk and Canny (1991); Bacow and Wheeler (1984); Burns and Ueberhorst (1988); for a review see: Fiorino (1990).

- A high degree of *reflection* on the different conditions and perspectives bearing on the threat in question
- An effective level of *communication* between participants concerning the different factual and value issues involved

The combination of the four discourse types forms the fabric of the envisioned political culture in food safety governance. Each of these discourses produces different types of outcomes that are fed into the next governance stage and enlighten the politically accountable decision makers. It is stressed that, while all participants should have equal rights in the deliberation processes themselves, the responsibility for the *final decision* lies with the *managers*.

To sum up: The General Framework advocates that public participation should be institutionalised, throughout the governance cycle, through the Internet Forum with open public access, and at the stages of framing and evaluation through an Interface Committee (in one of its two forms) bringing together assessors, managers, and key stakeholders. It further holds that a subset of food safety issues requires more extensive stakeholder and public engagement. The recommendation is that *procedurally* the intensity and form of engagement (participatory processes) be specified during the processes of framing and evaluation by the advocated Interface Committee in consideration of the given context and the overall socio-political climate. The Framework recommends, however, proceeding on the *preliminary assumption* that under the conditions of high levels of scientific uncertainty and socio-political ambiguity more extended participation which includes face-to-face participatory deliberation processes is of particular importance.

Chapter 8
Communication About Food Safety[1]

O. Renn

8.1 Introduction

In a thorough review of risk communication, William Leiss identified three phases in the evolution of risk communication practices (Leiss 1996: 85ff). The first phase of risk communication emphasized the necessity of conveying probabilistic thinking to the general public and to educate the laypersons to acknowledge and accept the risk management practices of the respective institutions. The most prominent instrument of risk communication in phase 1 was the application of risk comparisons. If anyone was willing to accept *x* fatalities as a result of voluntary activities, they should be obliged to accept another voluntary activity with less than *x* fatalities. However, this logic failed to convince audiences: people were unwilling to abstract from the context of risk-taking and the corresponding social conditions, and they also rejected the reliance on expected values as the only benchmarks for evaluating risks.

[1] There are two important points to make at the outset of the present chapter. First, throughout this chapter the term *risk communication* is used in a broad meaning to denote all types of communication about food safety threats. In our conceptual framework the term "risk" is defined in a strict sense, as referring to a situation where both the magnitudes of and the probabilities for a defined range of outcomes can be confidently quantified (see Sect. 1.1). In the present chapter, however, we have adopted the usual meaning of risk communication as a generic term to include any exchange of information dealing with the uncertain consequences of an event or an activity such as eating. In our terminology that will also include communication about uncertain or ambiguous food safety threats. A systematic differentiation between *food safety communication* (the term which is consistent throughout our overall conceptual framework) and *risk communication* (the term generally used in the existing body of literature which the chapter extensively refers to) would lead to confusion and would produce inconsistencies with the existing literature on risk communication.

The second point to mention is that this chapter had not been included in the early account of the General Framework which was put up for discussion in the workshop-based feedback and review process (cp. Chap. 11). Nor was it part of the revised version subjected for commenting (Dreyer et al. 2007a). The four invited commentaries (see Chap. 12) therefore do not relate to it. The chapter on risk communication was only added to the present volume, partly because several of the governance actors who provided us with a feedback missed a discussion on this topic and advised us to add a section on risk communication.

M. Dreyer and O. Renn (eds.), *Food Safety Governance*
DOI: 10.1007/978-3-540-69309-3_9, © Springer-Verlag Berlin Heidelberg 2009

When this attempt at communication failed, phase 2 was initiated. This emphasized persuasion and focused on public relations efforts to convince people that some of their behaviour was unacceptable (such as smoking and drinking) since it exposed them to high risk levels, whereas public worries and concerns about many technological and environmental risks (such as nuclear installations, liquid gas tanks, or food additives) were regarded as overcautious due to the absence of any significant risk level. This communication process resulted in some behavioural changes at the personal level: many people started to abandon unhealthy habits. However, it did not convince a majority of these people that the current risk management practices for most of the technological facilities and environmental risks were, indeed, the politically appropriate response to risk. The one-way communication process of conveying a message to the public in carefully crafted, persuasive language produced little effect. Most respondents were appalled by this approach or simply did not believe the message, regardless of how well it was packaged; this was also true for the area of food safety. The various food scares starting with BSE taught most people that the experts' assurances that all food items are safe, are often based on wishful thinking, and that uncertainties and ambiguities have been downplayed in order to avoid economic losses.

As a result of these communication problems, phase 3 evolved. This current phase of risk communication stresses a two-way communication process in which it is not only the members of the public who are expected to engage in a social learning process, but also the risk assessors and risk managers. The objective of this communication effort is to build up mutual trust by responding to the concerns of the public and relevant stakeholders. The ultimate goal of risk communication is to assist stakeholders and the general public in understanding the rationale of risk assessment results and risk management decisions, and to help them arrive at a balanced judgement that reflects the factual evidence about the matter at hand in relation to their own interests and values (OECD 2002). Good practices in risk communication help stakeholders and consumers to make informed choices about matters of concern to them and to create mutual trust.[2] Our approach to risk communication is inspired by the rationale of the third phase and is in line with the concept of *inclusive governance* that we have pursued throughout this book. The concern of the public about being well informed and included in the risk debate, highlights, according to the German sociologist Ulrich Beck, a gradual change within the predominant social conflict in modernity (Beck 1992; 2000). The primary conflict during the early twentieth century focused on the distribution of wealth among different social groups; after the Second World War, and particularly during the 1960s, the focus changed to the distribution of power in politics and economics. In more recent times, the major conflict has been about the distribution and the tolerability of risks for various social groups, regions and future generations.

[2] Cp. Hance et al. (1988); Lundgren (1994); Breakwell (2007: 130ff); Renn (2008: 240ff).

This shift of focus implies new forms of communication and collective decision-making between social groups and regulators, industry, civil society, and the public at large.[3]

Professionalization of risk analysis and institutionalization of risk communication are reinforced by the salient characteristics of risk phenomena in most risk arenas, including the one on food safety. The traditional process of decision-making in food safety relied on deterministic consequence analysis. Anticipating the most likely impacts of a decision and weighing the associated costs and benefits of different options, in terms of formal analysis or by "bootstrapping", had been the preferred methods of policy-making (Fischhoff et al. 1981). The questions of how to incorporate relative frequencies or probabilities within the decision process, how to cope with remaining uncertainties, and how to balance options with different compositions of magnitude and probability has become a major challenge for all food safety agencies (Zimmerman & Cantor 2004). A variety of strategies to cope with this new challenge has evolved over time. They include technocratic decision-making through expert committees or ignoring probabilistic information altogether (Löfstedt 2003: 423ff). The incorporation of probability assessments within decision-making requires new rationales for evaluating policy options and necessitates a revision of institutional routines (Freudenburg 1988). It is one major objective of the volume at hand to present new approaches to assessing and managing food safety threats under the different premises of full risk information, uncertainty, and ambiguity.

In addition, public perception of probabilities and risks varies considerably within professional analysis.[4] Whereas experts usually give equal weight to probabilities and magnitude of a given risk, the intuitive risk perception reflects higher concern for low-probability/high-consequence risks (cp. Covello 1983; Covello et al. 1988; Drottz-Sjöberg 2003: 16). Thus, risk communicators have to face the institutional problems of coping with the new challenge of stochastic reasoning and, at the same time, with the intrinsic conflict between the perspectives of the scientific community and the public in general (Rogers 1999; Kahlor et al. 2004; Breakwell 2007: 161). Both reasons justify the already established practice of highlighting risk communication in contrast to other forms of nutritional communication (Renn & Levine 1991).

As a consequence of this prominence, the interest of public institutions and academia in risk communication has grown considerably during the last decades. Risk communication has become a popular topic in the literature.[5] Although originally

---

[3]Cp. Luhmann (1989, 1993); Jungermann and Wiedemann (1995); IRGC (2005).

[4]Cp. Slovic (1987, 1992); Boholm (1998); Rohrmann and Renn (2000); Sjöberg (2000); Breakwell (2007); Renn (2008: 98ff).

[5]For overviews on the subject, see Chess et al. (1989); Covello et al. (1989); Leiss (1989); US National Research Council (1989); Atman et al. (1994); Morgan et al. (2001); Lundgren (1994); Gutteling and Wiegman (1996); UK Inter-Departmental Liaison Group on Risk Assessment (1998); Covello and Sandman (2001); Löfstedt (2001); OECD (2002); Drottz-Sjöberg (2003); STARC (2006); Breakwell (2007: 130ff); Renn (2008: 199ff).

conceptualized as a follow-up of risk perception studies, the work on risk communication has surpassed the limited boundaries of giving public relations advice for information programmes on risk, but extended its focus on the flow of information between subsystems of society.[6]

The following subsections deal with the concept of risk communication that lies at the heart of our food safety governance framework as it has been proposed in this volume. The second section explains the concept of risk communication and lists its major functions. The third section points out the requirements for risk communication at each stage of the food safety governance cycle. The fourth section describes the major risk communication approaches and instruments that could be used by communicating institutions. The fifth section explains the need for systematic evaluation of risk communication programs. The last section summarizes the results.

8.2 Definition and Objectives of Risk Communication

What is risk communication? The 1989 report on *Improving Risk Communication*, prepared by the Committee on Risk Perception and Communications of the US National Research Council, defined risk communication as:

> … an interactive process of exchange of information and opinion among individuals, groups and institutions. It involves multiple messages about the nature of risk and other messages, not strictly about risk, that express concerns, opinions or reactions to risk messages or to legal and institutional arrangements for risk management (US National Research Council 1989: 21).

Thus, risk communication fits into classic definitions of communication as a purposeful exchange of information between actors in society, based on shared meanings (DeFleur & Ball-Rokeach 1982: 133; Keeney & von Winterfeldt 1986). Purpose is required to distinguish messages from background noise in the communication channel. The term "message" implies that the informer intends to expose the target audience to a system of meaningful signals, which, in turn, may change their perception of the issue or their image of the sender (Jaeger et al. 2001: 129ff). Acoustic signals without any meaning do not constitute communication.

If one accepts the premise that risk communication implies an intentional transfer of information, one must specify what kind of intentions and goals are associated with most risk communication efforts. The literature offers different objectives for risk communication, usually centred on a risk management agency as the

[6] See Kasperson (1986: 275); Plough and Krimsky (1987); Luhmann (1990); Jasanoff (1993); Fischhoff (1995); Leiss (1996).

communicator and the public as target audience.[7] For the purpose of this essay, *objectives* can be divided into four general categories[8]:

- Ensure that all receivers of the message are able to understand its content and enhance their knowledge about the risk in question (*enlightenment function*).
- Establish a trustful relationship between the sender and the receiver of risk communication (*function of building up confidence in risk management*).
- Persuade the receivers of the message to change their attitude or their behaviour with respect to a specific cause or class of risk that relates, for example, to workers' protection, smoking habit or nutritional information on food (*function of inducing risk reduction through communication*).
- Provide the conditions for an effective stakeholder involvement on risk issues so that all affected parties can take part in a conflict-resolution process (*function of cooperative decision-making*).

These functions require specific types or forms of risk communication. In general, four different *forms of communication* can be distinguished (cp. Chess et al. 1989; Lundgren 1994; Renn 2008: 205ff):

- *Documentation:* This serves transparency. In a democratic society it is absolutely essential that, if the public cannot participate in the regulating process, people learn about the reasons why risk managers opted for one thing against another. Here it is of secondary importance whether this information can be intuitively grasped or understood by all. This situation is analogous to the information slips packaged with prescription drugs. Almost no one is able to understand them, save a few medically trained people. Nevertheless, these slips have important messages for the average patient, too. They illustrate that no information is being withheld (Jungermann et al. 1988).
- *Information:* Information serves to enlighten the communication partner. In contrast to documentation, information implies that the target group can grasp, realize and comprehend the meaning of the information.
- *Two-way communication or mutual dialogue:* This form of communication is aimed at two-way learning. Here, the issue is not a one-way street of informing someone, but an exchange of arguments, experiences, impressions and judgements.
- *Mutual decision-making and involvement:* In a pluralistic society, people expect to be adequately included, directly or indirectly, in decisions that concern their lives. The goal here is to ensure that the concerns of the stakeholders are represented in the decision-making process, and that the interests and values of those

[7] See similar classifications in: Chess et al. (1989); Hance et al. (1988); Morgan et al. (1992, 2001); De Marchi (1995); Mulligan et al. (1998); Sadar and Shull (2000); OECD (2002); Löfstedt (2003); Leiss (2004).

[8] See e.g. Covello et al. (1986: 172); Zimmerman (1987: 131ff); Kasperson and Palmlund (1988); US National Research Council (1989); Breakwell (2007:155ff).

who will later have to live with the risk effects will be taken up appropriately and integrated within the decision-making process.

Effective and inclusive risk communication simultaneously implements all four forms of communication. These four forms meet the various needs of diverse publics. Furthermore, these forms can be linked to the four functions mentioned above. Information and dialogue are the most appropriate means of achieving enlightenment; documentation and dialogue (in a conflict situation facilitated by mutual decision-making) of building up trust and of reducing risk, resolving conflict and encouraging mutual decision-making.

The following sections deal only with the first three functions: enlightenment, confidence-building and risk reduction by influencing behaviour. The fourth function of cooperative decision-making has been the main subject of Chap. 7 in this volume. However, the boundary between risk communication and participation is always fuzzy and difficult to draw.

8.3 Risk Communication Requirements for Each Stage of the Food Safety Governance Cycle

8.3.1 *Communication During Framing*

The first stage of the food safety governance cycle is dominated by the search for an appropriate frame under which the problem can be appraised and handled. The main task in term of risk communication is the assurance that all professionals involved in the subject area are, first, well informed and, second, enabled to provide feedback from their perspective.

The specific communication challenge in the framing stage involves overcoming organisational, internal communication barriers or communication barriers based on the application of different legal norms and institutions (Renn & Walker 2008). As we had discussed in Chap. 7, in some cases the same terms are used in different ways; in others, various risk assessment methods are applied to the same situation. In others, again, divergent justification forms are used or different statutory provisions apply (for instance with regard to the food item or the goal to be protected). For that reason it is essential for these different reference frameworks to become themselves a subject matter of communication even though insiders are completely familiar with these differences in the reference frameworks (BfR 2005). As soon as communication extends beyond departments (for example different panels of EFSA) or even risk assessment agencies (such as EFSA and the various national food safety organisations), the reference points which were seen as self-explanatory are, by no means, self-explanatory any more (Dreyer et al. 2006a). They must, therefore, be explicitly mentioned and communicated to all players involved. One way to assure a common understanding of the problem is to use the food safety governance framework that is advocated in this volume. Furthermore, control and feedback loops are to be envisaged at the interfaces between the public agencies to ensure that the intention behind the information reaches and is understood by the addressees. Within our framework, the Interface Committee is to be charged with this task of ensuring that all relevant actors are adequately addressed and ultimately included.

Often food safety issues also relate to chemical, animal protection, or ethical issues (De Jonge et al. 2007). The question, for example, of how to handle the problem of animal cloning for agricultural use has to be framed and re-framed in the language of human health, animal welfare, and ethical acceptability. Representatives from diverse disciplines and constituencies may come from institutions with varying territorial or functional competences (e.g. consumer protection, safety at work, ecotoxicology, etc.). The most important task here is the comparison and expert commentary of the data and the conclusions drawn from them. It is not about standardising but about avoiding inconsistencies, e.g. due to overlapping or missing competences. As a rule, the starting point of risk communication is a consensus on the common frame for further analysis and management.

When communicating with stakeholders and the public at large, risk communication at the framing stage should be inspired by two major goals: first, to ensure common understanding of the problem and second, to guarantee that alternative frames are collected and considered by the responsible authorities, in our framework the Interface Committee. The first goal is contingent on achieving a common understanding of the known terms and concepts that are familiar to all insiders involved. The central terms and concepts of assessment, evaluation and management should, wherever possible and legally admissible, be used for all external communication. It is particularly important that the terms and concepts clearly explain the degree of hazard, the overall context and the respective good to be protected. Whenever terms are to be used differently, an explanation of these differences should be given (BfR 2005). For instance, if the term "limit value" or "standard" has different meanings in different contexts then confusion and irritation are unavoidable. It should be made clear to the addressees that for formal, legal or contextual reasons, there has been a deviation from customary language use which is then, however, explained. Explanations of the key terms used are the first step towards achieving addressee-oriented processing of the material. Furthermore, risks must be presented in the overall context of risk–benefit analyses and the possibility of containing other risks by assuming a specific risk so as to position the risk and risk-containing measures in the overall context of the respective activity.

The second goal is to ensure sufficient opportunity for feedback (cp. Atman et al. 1994; Leiss 2004). In Chap. 7 we had already discussed formal ways of including stakeholders in the framing process. The representation of civil society in the Interface Committee and its inclusion through the Internet Forum (see Chaps. 6 and 7 for detailed information on these two proposed interface structures) are major contributions towards achieving this goal. In addition, one could organise systematic surveys or focus groups as a means of learning more about competing frames and to understand the concerns of society before the decision on the best risk assessment strategy is taken. A more refined concern assessment can then be performed during the assessment stage.

With respect to the general public, it is sufficient at this early stage to use media briefs or direct channels of communication (the Internet Forum) to inform all attentive audiences that the problem has been acknowledged and that the process of assessing, evaluating and managing the food safety threat has started. In addition, the information should contain the assurance that in the unlikely case that immanent dangers were to be detected, the fast route of prevention could be taken immediately. As we had explained in Sect. 2.2, we exclude from our analysis here the communication needs in case of a sudden crisis.[9]

This would require a chapter of its own.

8.3.2 *Communication During Assessment*

During the assessment phase communication is primarily directed to external scientists or experts from other public services, academia, and stakeholder groups. The main focus here is on an exchange of facts and arguments that are relevant for the characterisation of a risk or the assessment of the concerns. Handling and taking into account divergent views or divergent conclusions plays a major role in this stage. This is particularly true if the assessment reveals major uncertainties and ambiguities (Klinke & Renn 2002).

Communication during the assessment phase is primarily oriented towards collecting and appraising knowledge claims, i.e. critically examining the respective evidence, comparing interpretations of situations and giving adequate consideration to differing views. Communication focuses on the characterisation of a given food safety threat undertaken by the risk assessment agency and on the related consequences for assessment down to indications that are relevant in the later stages of evaluation and management. In this context, communication initially provided the basis for mutual understanding of each other's position and plausibly indicating how emerging differences in scientific opinions can be taken into account in the characterisation and assessment process (see OECD 2002; BfR 2005). The goal of communication here is the mutual inspection of evidence which is used as the basis for the respective assessments. The involvement of experts from external institutions should also help to procure further data on the topic, to collect and bundle different interpretations of the data and, finally, to arrive at a robust and reliable overall assessment of the physical risks, the associated uncertainties, and the accompanying concerns. The choice of experts should reflect the whole range of prevailing scientific opinions, cover all relevant disciplines, if possible, and give priority to independent individuals (Webler et al. 1995). As pointed out in Chap. 7, this requires an epistemic discourse with the major carriers of the relevant knowledge camps (for example by means of a Group Delphi as explained in Webler et al. 1991).

[9] For this see the review in Fearn-Banks (1996).

In addition to the involvement of external experts, the assessment stage also requires risk communication efforts targeted towards stakeholders and the general public. The main focus here is to inform all interested parties about the process of the assessment, the sources of information that the agency is using, and the timetable about when to expect the results. The major tool that we recommend for this purpose is the platform on assessment of the Internet Forum. More specifically, the communication in this phase should include (Ad hoc Commission 2003):

- Clear, timely and plausible documentation of all assessment processes and results, with information on the assessment methods and criteria used as well as on their factual and statutory bases;
- Information about the type of approach taken to assess the food safety threat (in our concept: *prevention*, *precautionary*, *concern-based*, *risk-based*; see Chap. 4);
- Information about the types of hazard and the corresponding risk by providing additional information on dose, exposure and contamination circumstances;
- Information on the relevant literature and other expert opinions;
- Information on how comments and tips from third parties are or have been taken over and processed;
- Information on participation and objection opportunities within the boundaries of the proposed epistemic discourse; and
- Setting up a "clearing house" on the Internet where interested users can access the latest information on the stage of the assessment and also ask their questions.

A third important element of communication in this stage of the process is adequate documentation. It is mandatory that the risk assessment agency documents all sources and refers to the data sets and references used. To the extent that it has input its own experience into the assessment, it is to indicate what this experience is based on. For instance reference can be made to one's own (not systematically evaluated) observations, anecdotal evidence, analogy conclusions or the conventions prevailing in the respective scientific community. It should be clear where scientifically validated evaluations and where the agency's *own* judgements have been adopted into the assessments. All sources and conventions used in the assessment process should be publicly documented, most preferably on the assessment platform of the proposed Internet Forum.

8.3.3 Communication During Evaluation

Food safety agencies such as EFSA are frequently accused of not taking due account of or even ignoring the diversity of values and lifestyles, which means that there is no consensus on which risks are acceptable or tolerable (see Löfstedt 2005; Bandle 2007). Furthermore, criticism is uttered regarding inadequate plausibility when it comes to specifying protection goals,[10] the level of protection, or priorities in conjunction with competing protection goals. During the stage of evaluation, the focus is on different values and weighing up criteria which may vary considerably in a pluralistic society. What is needed is the timely involvement of the representatives of stakeholders in this phase (Bunting et al. 2007). Conflicts would be less severe in this stage, if the main stakeholders were to be involved during the framing stage, i.e. when establishing the protection goals, the level of protection and the setting of priorities. Therefore, we advocate formalising participation of key stakeholders at these two stages through the Interface Committee.

Ideally, all stakeholders who feel affected by the tolerability or acceptability judgement would be included in the evaluation stage. The Internet Forum can play a major role here for involving a wide diversity of social groups. An explicit invitation to take part in the communication (a "top-down" logic of involvement, cp. Sect. 6.4.3.1) may be addressed to the risk initiators, delegates of organised interest groups such as representatives of industry, unions, associations, nature conservation organizations, or autonomous players (WHO/FAO, government representatives on all governance levels, specialist agencies, political parties).

Of course, a decision must be taken in advance about which groups and respective goals are to be involved at this stage. As a general rule, as many groups as possible should be given an opportunity to speak in order to permit the entire range of arguments, concerns, worries and interests to be presented. However, for the sake of efficiency, depending on the degree of uncertainty and ambiguity, the number of participants may be limited. In Chap. 7 this point has been discussed in more detail.

The main purpose of the communication with stakeholder groups at this stage of the process is to ensure that all relevant values and arguments for making a prudent judgment are being considered and deliberated. More specifically, the process should be based on (OECD 2002):

- A mutual understanding of diverse points of view and their interpretation(s);
- A written or face-to-face exchange of interpretations, criteria for evaluation, and input for balancing benefits and risks;
- An integration of the concerns, worries and interests of the stakeholder representatives into the evaluation process;
- An effort to reach a joint agreement on the further procedures for managing the food safety threat.

A second major goal of communication in the evaluation stage is to improve mutual trust and credibility among all actors concerned (see Covello 1992; Renn 2007). To serve this goal, communication should focus on the exchange of all

[10] For example in the case of listeria, where cheese farmers in southern France do not understand the judgement of food safety agencies to ban cheese produced by non-pasteurized milk (Knight et al. 2008: 209ff).

arguments, interpretations, trade-offs and assessment results that have gone into the final judgement on tolerability and acceptability. Communication should facilitate the understanding of the various standpoints and backgrounds that have been incorporated into the decision. Ideally, risk communication can identify how the concerns and interests of all the stakeholders have been considered and processed at the evaluation stage (Siegrist et al. 2000) and, in our framework, provide this information via the Internet Forum. In this context, possible disadvantages for individual groups resulting from the weighing-up process would be presented in a transparent manner and plausible explanations given as to why they were deemed unavoidable.

Public information about the process and the results of the evaluation stage, i.e. the tolerability/acceptability judgment, should be guided by the following principles (BfR 2005):

- The communicator (e.g. the food safety agency or the Interface Committee that we propose) should explain the weighing-up process for the tolerability/acceptability judgement and stress its willingness to accept proposals from other social players about this judgement.
- The communicator should provide a clear, logical justification for the trade-offs that she/he has used to make the tolerability/acceptability decision.
- The communicator should always be accessible through an Internet link for collecting feedback from the general public, i.e., in our concept, the platform on evaluation of the Internet Forum.
- The communicator should be available for discussions with media representatives about this evaluation process.

When communicating evaluation results to the media or different public audiences (such as consumers, retailers, food critics, etc.), the communicating institution should keep the concrete socio-political context of the target group in mind as the risk information should be directly linked to the life circumstances of the participants. It is essential that the communicator orients him- or herself towards the addressees' interests taking into account their needs and concerns when presenting the results of the evaluation process. Statements reflecting expert assessment of risks and concerns have to be cast in a format that the target audience is able to understand and digest. Technical terms should be avoided if possible, or only those terms should be used which are essential to understand the statement. Central statements should be rendered clearer through illustrations from the area of the addressees. Complex situations should be presented graphically if possible.

With respect to information content, the communication to the various publics should be guided by the following design criteria (BfR 2005):

- The communicator should communicate all risk-relevant findings and all arguments that were used in the trade-off procedure to ensure full transparency.
- The communicator should explain the quality of the knowledge basis and stress that more exact and improved results are to be expected in future

through further research. The communicator should indicate a deadline and responsibilities for the expected research results.

- The evaluation process should be related to the experiences of the receivers. Consideration should be given to the differences between the individuals or groups concerned within pluralistic societies or between various cultural subgroups (for instance groups with special diets).
- The communicator should pass on the available scientific findings, experiences, assumptions, or presumptions as well as the judgements arrived at, assessments, interpretations or conclusions in a way enabling the addressees to make up their own minds.
- The communicator should reveal where there is clear scientific evidence and where there are scientific uncertainties necessitating the precautionary approach of assessment.
- The communicator should also refer to risk perception variables and the results of the concern assessment to the degree that they have influenced the evaluation process (Breakwell 2007). Such variables include (see Slovic 1992; Sjöberg 2000; Rohrmann & Renn 2000):
  - Personal or institutional control opportunities;
  - Maximum scale of disaster (what can happen in the worst case?);
  - Sense-related perceptibility of the risk (can a consumer detect the risk by own means?);
  - Perceived opportunities for self-protection;
  - Perceived distribution justice (are specifically vulnerable groups at risk?);
  - Perceived benefits.

Risk communication during evaluation should ensure full transparency about all implicitly applied or integrated value judgements. If it is scientifically possible, probabilities should be quantified (for instance: four expected cases of disease when 10,000 people are exposed to a specific food substance). In addition, the communication should include information about the reference that was used to determine the desired protection target (if specified), the reasons for the choice of safety factors, NOAEL[11] levels or other normative conclusions, and a characterisation of the remaining uncertainties and residual risks.

8.3.4 Communication During Management

The management stage specifies the measures of how to deal with a given food safety threat. Such measures include licensing procedures, standards, economic incentives or disincentives, labelling, or technical specifications (IRGC 2005).

[11] *N*o *o*bservable *a*dverse *e*ffect *l*evel (NOAEL).

Depending on the type and nature of the measure considered by the managers, different target audiences are directly or indirectly affected. For example, industry is the prime communication partner if the measures are directed towards specifying product composition, processing requirements, or concentration standards. In this case, communication must demonstrate that the chosen measures are (Renn 2007):

- Effective to meet the desired safety goal;
- Efficient in the sense that no other measure could meet the same purpose with fewer costs in terms of money and other resource investments;
- Fair with respect to those who bear the costs and those who will enjoy the benefits;
- Congruent with the values and ethical norms that apply to this food safety threat; and
- Feasible and operational with respect to the legal norms, technical requirements and political priorities.

The tasks of risk communication vary, if the measures are directed towards the final consumer. These measures include product labelling, food advisories, or educational programs. The German Ad Hoc-Commission on the Harmonization of Risk Standards (2003) has coined a special term to illustrate the goal of risk communication when directed towards informing the consumer about risk reduction measures. The term they created is *risk judgement sovereignty*, meaning that all risk communication efforts should allow for every interested citizen to be able to make a personal assessment of the respective risks – in line with his/her own evaluation criteria, personal preferences and/or with ethical criteria he/she deems appropriate for society – and always provided that the citizen understands the proven impact of a product, the remaining uncertainties and the justifiable interpretation scope.

In this context, communication is an open process of the mutual comparison of information and arguments. It should not be the aim of risk communication to persuade individuals to treat risks in a standardized way, but to inform consumers of the options they have to reduce their risks in accordance with their own preference structure (BfR 2005). For this purpose, they need to understand the implications of consuming a risky product, understand the probabilities and uncertainties surrounding these impacts, and to develop the appropriate judgmental capability to integrate this knowledge with their own values and preferences in order to form a balanced judgment about what to do.

Risk management agencies (such as DG SANCO on the European level) have therefore the responsibility to inform the consumers about the mandatory measures (such as concentration limits) imposed on the food producers and suppliers, but in particular about the voluntary options that individuals have in reducing their overall risk from consuming food items. For this purpose, communication should be guided by the following principles (see OECD 2002; Renn 2008: 249ff):

- Communication should include simple, clear messages and be appropriate despite the complex nature of the subject. Frequently, it helps to present the simple important messages at the beginning of the text and to deal with the more complex elements at the end of the message. Very interested readers or listeners are very willing to digest the entire text of the message; people who only have a superficial interest in the subject will feel that they have already obtained sufficient information from the introductory sentences.

- Communication should be flexibly adapted to different situations by employing the most suitable methods. Depending on the communication context, other media or transmission channels are to be selected.
- The material for communication should be complete and should contain all relevant information on the risk and on the risk reduction options. It should provide references for more in-depth information if consumers are willing to invest more time.
- The material should be well illustrated and should provide intuitive access to the scientific foundations, the statutory provisions, the scope for action and the chosen risk reduction measures.
- The communication should explain and illustrate all behavioural measures that could be applied for the purpose of risk reduction or avoidance.
- The text of the communication should also be aimed at including particularly sensitive groups to an extent which makes sense within the framework of the risk involved. What is particularly important is information on suspected effects in infants, children, senior citizens, and the chronically sick.
- The communication should admit any remaining uncertainties and demonstrate that the risk management agency can adopt a precautionary stance when there is sufficient suspicion of potential damage (explanation of the precautionary principle).
- The communication text should avoid bureaucratic and legal jargon but should pay attention to the possible legal implications of a statement.
- The risk management agency should assure the public that it is available at any time to answer further questions or accept comments, and should give the names of the corresponding contacts.

When it comes to solving complex risk management problems and communicating them to a wider public, what is normally needed is a mixture of various but mutually interacting control and communication methods. Neither the experts with their technical understanding nor the stakeholders with their values can claim sole legitimacy for justifying management measures. If these measures are highly controversial, it is vital to communicate about them, in particular through the platform on management of the Internet Forum, but this is not sufficient. In these cases, as has been stressed in Chap. 7, innovative methods of involving stakeholders and the general public are particularly needed. Novel tools such as citizens' consensus conferences, citizens' fora and future workshops can be used in addition to the classic instruments such as hearings, panel discussions, and public group meetings (see Rowe & Frewer 2000; Renn 2008: 330ff). A participatory approach is particularly recommended if the management measures affect the interests or values of the individuals or groups concerned, to a major degree, and if the costs and benefits of these measures are very unevenly distributed throughout the population.

8.4 Communication Tools

Which tools can be assigned to communicative and dialogue-driven procedures? There are three basic types of communication tools (see Wiedemann & Schütz 2000; Renn et al. 2005) (Fig. 8.1):

Information-based tools: These encompass all forms of communication oriented towards the communicating body informing the target group(s). Feedback or two-way communication is not envisaged. This type of communication should be selected when the group of addressees is very large and communication can be ensured through the mere transmission of information. Frequently information-based tools are suitable for the preparation or subsequent processing of dialogue or participation-based tools. The information-based tools include brochures and other written material, newspapers, classical public relations (such as press releases and radio interviews), websites, events, etc.

Dialogue-based tools include two-way communication with the addressees of communication, but without the addressees being given the opportunity to play an active role in the design, assessment or implementation of decisions and measures. Dialogue is, therefore, restricted to questions and answers, explanations and questions, the sounding out of opinions and judgements as well as reciprocal information. Dialogue-based tools include brochures with a return coupon, opinion polls, lectures, panel discussions, discussion rounds, Internet with feedback, chat-rooms, dialogue-based events, and open days for visitors.

Participation-based tools: The participation-based tools differ from the dialogue procedures in that they directly or indirectly integrate the concerns of the addressees into the decision-making process. Here the boundary between dialogue and participation is often fluid. The participation methods can be classified in three groups:

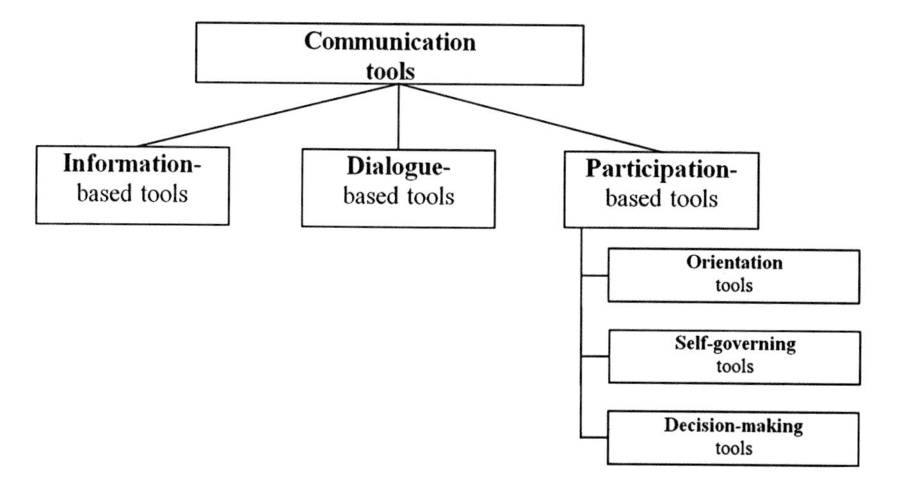

Fig. 8.1 Communication tools (adapted from BfR 2005)

- *Generating ideas and orientation:* The orientation tools are designed to allow the groups concerned to help orient decision-makers without influencing them directly. The goal of orientation is for decision-makers to get to know and understand the concerns of the groups. Furthermore, some tools aim at discussing joint options with the group representatives and at reflecting together on the advantages and disadvantages of each option. Orientation tools include the Internet Forum that we advocate (it can act as an entry point into the governance process at all four major governance stages of a diversity of viewpoints) and also hearings, non-binding round tables, citizens' assemblies, open space conferences, and focus groups.
- *Self-commitment and self-governance:* These tools are about co-ordinating actions which are carried out and implemented by the players themselves. The political decision-makers may provide stimulus or an organisational platform for this discourse. For instance measures may emerge which are in the interests of both groups. The self-governance tools include working groups, future workshops, open space conferences (also suitable for espistemic discourses), round tables.
- *Recommending a decision:* The decision-making tools involve the concrete preparation of a political (i.e. collectively binding) decision in the form of management recommendations or the decision itself. Discourses of this kind are appropriate when specific groups or representatives of the general public are to be directly involved in the decision-making process. In some cases participation of this kind is stipulated by law or is used consciously by political decision-makers in order to take the concerns of those affected by the decision into account and to secure their positive response to the decision. The Interface Advisory Committee that we recommend (see Chap. 6) can be described as such a tool, as it advises the managers on the terms of reference as well as on the evaluation of food safety threats, the latter being an important input into the management decision-making. Other tools suited for this method are, for instance, round tables, co-operative planning rounds, citizens' fora, consensus conferences and mediation (in the case of conflicts).

The selection of the appropriate tools depends on the risk issue and the communicational function that the communicator intends to meet (Rowe & Frewer 2000). Regardless of what tool is being applied, the targeted audience expects high commitment and excellent performance. The addressees have high expectations of risk communication. The starting position for communicating to the public is not easy. The staff of risk assessment or management agencies faces increasingly emotional reactions, growing pressure to justify themselves and to offer objective insights and unbiased advice (Löfstedt 2005). Many risk assessors and managers are forced into a reactive role and must deal with ongoing dissent, uncertainties and a widening scope for action in the assessment as well as management process (Luhmann 1980). Communication programmes and internal structures must adapt to these new conditions (Wiedemann & Schütz 2000). How can institutions succeed in this?

Risk communication cannot be done "in passing". It needs to be integrated into the organisational structure of the institutions involved in risk assessment or management (Ad hoc Commission 2003). It is essential for risk communication to be firmly anchored in the risk assessment and risk management institutions, and for risk communication experts to be recruited to the teams of risk analysts and risk managers. Moreover, the participating scientific risk analysts and risk managers must be equipped with the communication skills needed to exchange their approaches and results among themselves and with the other players and to present them to the general public in an understandable and plausible way. Risk communication must be seen as an *integral component* of the entire regulation process starting in the preliminary phase of framing right through to the implementation of measures. All risk communication efforts should be timely and comprehensive, and reflect the concerns of the targeted audiences.

8.5 Evaluation of Risk Communication

The actual implementation of risk communication decides, in the final instance, on the achievement of the desired goals. It is, therefore, essential to evaluate risk communication in order to assess its effectiveness (see Rohrmann 1992, 1995; Renn et al. 2005). The evaluations should be an integral part of any risk communication programme. During evaluation the contents, procedures and consequences (results and effects) of risk communication activities are being scientifically assessed using specific criteria with regard to the previously specified goals. The risk communication programme on food safety touches on highly relevant social topics such as health protection and central safety needs. For that reason alone, there is a need for an evaluation of the communication efforts if success is not to be left to chance. Common sense or subjective individual opinions are not enough. Intuitive efficiency assessments are misleading because of selective and mood-driven perceptions (De Jonge et al. 2007). For that reason, a systematic and empirically backed approach is essential. In order to ensure the implementation of the risk communication models, there is a need not only for scientifically backed but also ongoing evaluation along the lines of permanent quality control.

Not only can an evaluation highlight whether and, if so, to what degree the desired goals have been achieved but also which elements in the risk communication programme have contributed to achieving, or failing to achieve, the desired goals (Bostrom et al. 1994). A strength–weakness analysis of this kind can also be used when a decision has to be taken about continuing or abandoning the programme or looking for an alternative. An evaluation also serves to justify the costs and resources needed for a comprehensive risk communication activity, and to peg out the foundations for efficiency considerations (Goldschmidt et al. 2008). At the same time, a report must be given about the extent to which the communication corresponded to the needs of the targeted audiences. In this way, evaluation creates the empirical foundations for optimising the risk communication programme by providing a basis for a decision on the setting of priorities within the diverse range of possible combinations of tools.

The evaluation should be conducted by external experts or by trained staff members who are experienced in putting together questionnaires, carrying out surveys, dealing with preconceived ideas or evaluating different data material (OECD 2002). Ideally, the assessment of the evaluation does not only look at the actual results but also at its unintended effects. Here the evaluation can concentrate both on the observation of internal and external effects. Good evaluation need not necessarily be complicated. What is important for compliance with a given cost framework are clear ideas about the goals to be achieved by the evaluation.

An evaluation is interested in future-oriented, constructive recommendations for improvements and not in destructive criticism of the past. What is necessary here, is the willingness of all stakeholders to submit their performance to critical observation. An evaluation should be announced in advance to all players. It must be clear to everyone as to who is to be evaluated, what is to be evaluated, and the scale of the evaluation.

There are numerous ways to perform an evaluation.[12] Among the most popular are:

- *Preliminary analysis, pre-test, focus group:* Here the material or the evaluation procedure of the future evaluation programme is tried out in a test group (focus group). In simulations and role plays the effect of the "key message" can be tested. A preliminary test reveals whether there are blockades in the flow of information and how the material can be improved. This can prevent "unpleasant" surprises. The method is effective and highly efficient, and should be an essential part of all risk communication activities.
- *Systematic feedback:* Systematic feedback involves obtaining feedback on risk communication activities directly from, if possible, all those concerned. In the case of oral communication, assessment sheets and short questionnaires can be distributed or, in the case of written communication, response forms can be attached (e.g. performance check). This method is extremely cost-effective, user-friendly and the results are rapidly available. However, the questions must be carefully couched; there should not be too many and they should permit clear answers.
- *Experimental design:* The classical form of experiment design is the comparison test with a control group who were not "exposed" to any risk communication activity (stimulus). This test has the advantage of being able to measure the effects of risk communication activities directly and without any possible third-party influential factors. However, the time and costs involved are considerable.
- *Surveys and interviews:* A representative selection of all the people directly concerned are being questioned using a standardized or open questionnaire. From the angle of risk communication this does not so much entail surveys by

[12] See Rohrmann (1992, 1995); Bostrom et al. (1994); OECD (2002); Renn et al. (2005); Renn (2008: 318ff).

opinion-poll institutes, intended for the overall population, as it entails a survey of the targeted audiences. The interviews can be recorded and then evaluated from the qualitative angle (this takes a lot of time, however). The interview offers the advantage of giving the interview partners an opportunity to immediately clarify unclear questions and to identify individual priorities.
- *Chat analysis:* Internet chat rooms can be used for various purposes in order to pass on information to consumers, to enter into a dialogue with them and to collect information about one's own performance. On the Internet, participants communicate directly and anonymously with each other in real time, like in a forum. In addition to the contents, the written dialogue provides further assessment aids. Software programs permit rapid and comprehensive analysis of the arguments used and the profiles of the participants. The results obtained are limited in terms of their impact as the participants merely represent a specific participant circle (computer users). But chat analyses provide a rapid and relatively low cost opportunity for assessment by communication partners.

8.6 Conclusions

The communication process can be compared to a free market system in which goods are produced, transported, purchased, and consumed. In the long run, most of the good products will find their market niche, whereas the majority of bad products will eventually fail to meet the market test. Similarly, messages containing important information are more likely to reach their destination; but many trials may be needed to ensure this success. In addition, packaging can help to sell the message faster and to overcome obstacles on the way from the source via the transmitter to the final receiver. The package can help, if the message is worth transmitting; but even the best package will fail in the end, if the message is meagre, dishonest, or simply irrelevant. Almost any risk communication study is quick to point out that risk communication is not a public relations problem (see Gray et al. 1998; Bennet & Calman 1999). Advertisement and packaging of messages can help improve risk communication; but they cannot overcome the problems of public distrust in risk management institutions or cope with the incapability of the current risk arena to produce rational and consistent risk policies. The potential remedies to these two problems are: better performance of all institutions dealing with or regulating risks, and restructuring the risk governance cycle to meet the requirements of effective and transparent risk handling.

With regard to a good performance record as a prerequisite for credibility, many risk management institutions face the problem that their specific task is not well understood and that public expectations do not match the mandate or the scope of management options available to these institutions. This is certainly not unique to risk management agencies. Lipset and Schneider (1983) found out that elites in the US complain regularly about the ignorance and misconceptions of the public with respect to their mandate and performance. Regardless of whether this claim is true, a clear gap separates the self-perception of most institutions and their public perception. This is specifically prevalent in the risk arena because health and environment top the concerns of the public, and because the stochastic nature of risk impedes an unambiguous evaluation of management success or failure (Johnson 1993).

In spite of these difficulties, careful management, openness to public demands and continuous effort to communicate are important conditions for gaining trustworthiness and competence (see OECD 2002; Renn et al. 2005). They cannot guarantee the success; but they make success more likely. Therefore, the *first principle of good risk communication practice is to start with a critical review of one's own performance.* Is the performance good enough to justify public trust? Are mechanisms in place that help discern the needs and requests of stakeholders and the general public? Is a two-way communication programme implemented? Is the communication honest, clear, comprehensive, and timely? Have all requirements for a transparent and accountable risk governance structure be met?

The second most important principle of risk communication refers to its position in the risk management process. Many risk managers believe that risk communication starts after the management process is completed. Our food safety governance framework suggests, however, that risk communication must be an ongoing activity during all governance stages, i.e. framing, assessment, evaluation, and management.[13] *Therefore, the second principle of good risk communication is to design an integrative food safety governance and communication programme ensuring a continuous effort of communicating with the most important stakeholders and the consumers from the framing to the management stage.* In the early phases of framing – the identification of the problem and the choice of the appropriate objectives and criteria – risk communication needs to address issues such as the proper institutional umbrella under which the problems fit, the plurality of concepts and reference points to deal with the problem, the choice of methods and techniques to identify problems and to ensure public protection, and the setting of priorities in dealing with many problems at the same time. During the assessment stage, the protocols for prevention, precautionary, risk and concern assessment need to be communicated to those who have an interest in the assessment process. Feedback is also required in terms of collecting experiential and local knowledge about the problem at hand. In later stages, i.e. during evaluation and management, the rationale for making trade-offs between conflicting objectives, the targeted level of protection as well as the selection of management options need addressing. Questions in this context are: how do managers detect problems before it is too late? What criteria are being used for evaluating food safety threats? How is the decision process designed to accomplish an optimal trade-off between economic, environmental and public health objectives?

If these questions can be positively answered, the designing of communication can be further optimised. *The third principle of good risk communication practice is to tailor communication according to the needs of the targeted audience and not*

[13] Similar in: Leiss and Chociolco (1994); Jungermann and Wiedemann (1995); Leiss (1996).

to the needs of the information source. Information should match public expectations. As trivial as this appears at first glance, it is one of the most violated principles in risk communication. Targeting the message to the needs of the audiences requires more than a good intuition what the public allegedly needs to know. Targeted risk communication depends on state-of-the-art surveys about the information needs and the perceptions of the targeted audience (Fischhoff 1995). It is not sufficient to confine the communication process to a discussion of probabilities and consequences. Communication should include aspects such as whether the exposure is voluntary, what possibilities exist to exert personal control (or if that is not feasible, what institutions can fill that gap and monitor and control risks on behalf of the public), how the risk and its consequences are being managed, and how catastrophic events can be avoided. Risk communication is particularly difficult if risks are invisible to the consumer and may cause negative health effects after a long incubation time (Renn 2008: 115ff). These risks are particularly frightening for the consumer: they are associated with involuntariness, delayed effects, inability to be sensed by human organs, lack of control and unfamiliarity. To address these negative risk characteristics, it may be helpful to point to functional equivalents of these characteristics in a broader societal context. Potential equivalents are the assurance of a democratic decision-making process to counteract the impression of involuntariness and, as a replacement for personal control, the independence and impartiality of operating and regulating agencies. This may produce trust in their capability to monitor food items on the shelves, check composition and durability of goods and intervene if safety in the risk-producing facility is not managed properly (Barr 1996). In addition, unfamiliarity can partially be compensated for by better functional knowledge about the risk and the associated technology.

The fourth principle of good risk communication practice is to adjust and modify one's communication programme as a result of an organized effort to collect feedback and to sense changes in values and preferences. Many successful programmes of the past have turned out to be inappropriate in addressing today's audience. Constant adjustment requires efforts to collect systematic feedback from the community, the relevant stakeholders and the general public. This calls for a continuous evaluation programme.

Even if all these suggestions are followed, risk communication may not work (Trettin & Musham 2000). External influences, the overall climate of distrust, past management failures and specific incidents can transform risk communication into a never-ending frustration. This frustration – so familiar to most risk managers – is an indication of the need for a more fundamental risk discourse. The ultimate goal of a risk communication programme is not, to ensure that everyone in the audience readily accepts and believes all of the information given, but to enable the receivers to process this information in order to form a well-balanced judgement in accordance with the factual evidence, the arguments of all sides, and their own interests and preferences. To accomplish this goal, a risk communication programme is needed to provide the necessary qualifications to all participants and to empower them to be equal partners in making decisions about risk. We consider the setting up of the Internet Forum, closely linked to the Interface Committee, a major element and tool of such a programme.

Chapter 9
Implementation of the General Framework: Genetically Modified (Cry1Ab) Maize Case Study[1]

A. Ely

9.1 Introduction

This chapter works through the case of placing on the market for consumption as food (not cultivation or feed) of *Bacillus thuringiensis* (Bt) Cry1Ab transgenic *Zea mays* in order to demonstrate how the food safety governance framework introduced in the earlier chapters of this book could be implemented. It does not make prescriptive judgements regarding decisions that the respective institutions should make (e.g. around terms of reference, screening criteria or assessment outcomes), however it explains the mechanisms through which each of these stages would be executed, suggests possible results at each of these junctures and explains the potential consequences in terms of subsequent stages in the governance framework.

Bt maize is among the first generation of genetically modified foods that were submitted for regulatory appraisal within the European Union (as early as 1994[2]), and several events have received food safety clearance from EFSA. It is maize that has been engineered to express insecticidal toxins from the bacteria *B. thuringiensis*. Cry1Ab is a type of toxin that targets certain Lepidopteran pests (butterflies and moths). The example is reminiscent of past product notifications for Bt176, Mon810 and Bt11 under the Deliberate Release Directive 90/220 or the extension to include Bt11 sweet maize (to the Netherlands) under the Novel Foods Directive 258/97, as well as subsequent applications (through various legal procedures) for stacked varieties derived from the aforementioned events. In addition, brief reference will be made to the discovery in December 2004 that Bt11 maize planted in

---

[1] As with Chap. 8, the present chapter had not been included in the early account of the General Framework that was put up for discussion in the workshop-based feedback and review process (cp. Chap. 11). Nor was it part of the revised version which was subjected for comments (Dreyer et al. 2007a). The four invited commentaries (see Chap. 13) therefore do not relate to it. As with the preceding chapter on risk communication, the case study chapter was only added to the present volume. This was mainly a response to the feedback of several of the governance actors engaged in the framework's development who requested an illustration of the working of the governance framework through a case study.

[2] Bt176 was submitted to the French competent authority in 1994 under Part C of Directive 1990/220 (notification C/F/94–11–03 originally submitted by Ciba Semences).

M. Dreyer and O. Renn (eds.), *Food Safety Governance*,
DOI: 10.1007/978-3-540-69309-3_10, © Springer-Verlag Berlin Heidelberg 2009

2001–2004 was contaminated with Bt10, a different Bt maize event (MacIlwain 2005). Had this discovery been the result of illness or mortality in consumers, this historical example might be thought of as a 'food-scare' event (of the type that the food safety governance framework was not formulated to deal with). However, in actual fact the contamination was discovered through laboratory research by the firm involved (Syngenta), and thus might more accurately fit within the monitoring stage of management. Here the framework illustrated in this case study could apply, with the new information in monitoring being fed into the framing stage for later screening/assessment by EFSA.

Aspects of each of these historical cases will be mentioned in the chapter; however, in order to demonstrate the framework as clearly as possible, the hypothetical case study presented here will draw upon recent scientific debates as if a new Cry1Ab event were submitted for human food use under the contemporary legal framework (i.e. Regulation 1829/2003 on genetically modified food and feed). As will be demonstrated in the later sections regarding ambiguity, it is possible that the concerns of some stakeholders will necessarily refer to issues beyond human food safety (namely the use of Bt maize in animal feed and for cultivation), leading to their brief mention in the hypothetical case study presented here.

This case study will run through each of the stages in the proposed governance framework outlined in Chaps. 3–5 individually. It should be remembered that the framework is flexible and able to respond to requirements for feedback or repetition of certain activities (e.g. referral back to EFSA following the identification during evaluation of a salient issue that was previously neglected in assessment). For the sake of simplicity, a brief overview of framing, assessment, evaluation and management are provided, without detailing all of the instances where this sort of feedback might occur.

9.2 Framing

9.2.1 Review

In the case of Bt maize, review refers not only to the adaptation and improvement of legal and institutional contexts within which the product is handled within the European Union, but also to the international environment which acts to shape the European context. In both cases, informal conventions and dominant practices as well as codified legal texts, are significant. At international level these might include the agreements of the World Trade Organisation (including the Sanitary and Phytosanitary Agreement as well as the Agreement on Technical Barriers to Trade), the non-legally binding OECD (2003) consensus document on maize (*Z. mays*) (OECD) or the implications of the Cartagena Protocol on Biosafety for the EU's supply to export markets.

At the EU level, the General Food Law[3] acts as a basis for the governance of food safety and would apply in all cases where new food products are to be put onto the market. In addition, certain existing legal instruments are explicit in the ways in which they frame assessments. For example, Decision 1829/2003 on genetically modified (GM) foods refers to the principles for assessment set out in Annex II of Directive 2001/18/EC. In so doing it stipulates the types of studies that must be carried out by specifying the information required for submission to EFSA in Annexes III of Directive 2001/18EC, and also makes demands on those marketing the foods by specifying the types of labelling requirements laid out in Annex IV of Directive 2001/18EC, as amended by Regulation 1830/2003. As previously explained in Sect. 3.2, review involves the adaptation of these legal frameworks not only in response to developments in scientific understanding (based in part on monitoring the effectiveness and consequences of existing management measures, but also on emerging upstream/basic research findings) but also to shifting socio-political, legal and institutional contexts at national, EU and supranational levels. With regard to the setting of risk assessment policy around Bt maize, the issues at stake might include, but not be limited to:

- The level of proof of safety required (with the burden placed on seed firms wishing to introduce Bt maize to the market) (the chosen level of safety)
- The attributes to be tested for, whether they be allergenicity, toxicity, nutrition or other
- The time-scale (number of generations) and diversity and representativeness of samples to be tested in assessing each of these attributes
- The need to investigate any locale-specific, culturally sensitive or other distributional issues that might be linked to the product
- The alternative food options against which the *pros* and *cons* of Cry1Ab Bt maize should be assessed and evaluated
- The range of options available, beyond mere approval or disapproval of the product

Under the General Framework, the process of review at the EU level would strive to be conducted openly and systematically, not only focusing on the types of scientific information required (under annexes) in the above-mentioned Directives and Regulations, but also importantly in specifying the criteria against which each case would be screened in order to inform the most appropriate form of assessment. Here the Interface Committee, discussed in detail in Chap. 6, would play the vital role of facilitating the participation of a broad range of stakeholders in setting these criteria. The criteria would be specified and continuously reviewed in order to take into account evolving scientific knowledge and concerns around genetic modification and Bt maize in particular. Assessors, managers and other stakeholders could, through the Interface Committee, draw on the general screening criteria described in Sect. 4.2 (and later in this chapter) and amend them to focus on the case at hand

[3] Regulation (EC) No 178/2002 (*OJ* 2002, L31/1) as amended by Regulation (EC) No 1642/2003 (*OJ* 2003, L 245/4).

(for example through further narrowing them to cater for genetically modified crops in general or insect-resistant crops, if such specificity was felt to be warranted). Although screening represents a departure from the existing regulatory procedures currently in place, the criteria might be based upon the information required by the annexes to Directive 2001/18/EC, and their setting should therefore not be overly onerous. Some of the potential challenges that screening, as one of the most innovative components of the General Framework elucidated here, could raise for EFSA are discussed in Chap. 6. It is clear that the introduction of this component would require an iterative learning process, within which transparency and accountability will be key aspects. These principles are addressed by explicitly providing for the involvement of assessors, managers and other stakeholders during the process of review (including the setting of the screening criteria).

9.2.2 *Referral*

Referral is the act of referring a question or product notification to EFSA for assessment, and can be carried out by a number of actors within the EU (the Commission, a Member State or EFSA itself). In the case at hand, it would involve the national competent authority of a Member State drawing upon the legal provisions mentioned in the section above (usually Regulation 1829/2003, Art. 5, Sect. 2a) to refer the specific case of Bt Cry1Ab *Z. mays* to EFSA. The discovery in 2004 that 4 years of Bt11 field maize harvests in the USA had been contaminated with Bt10 (a different transgenic maize event, which includes the *bla* antibiotic resistance marker) was not a conventional referral as covered in the General Framework. In this case it was the firm Syngenta that discovered the contamination when it upgraded its quality assurance practices, and notified US regulators themselves. EU member states and the European Commission were informed of this contamination on 22 March 2005, when an article appeared in Nature (UK Advisory Committee on Novel Foods and Processes 2005; MacIlwain 2005).

In an application for approval of a new Cry1Ab Bt maize product under Regulation 1829/2003 on genetically modified food and feed, the information available would be forwarded to EFSA for screening by a Member State, which (largely because the activity of screening is carried out by EFSA) is itself treated within the General Framework as a component of assessment. In order to simplify the narrative of this case study, screening will be dealt with here (rather than in the following assessment section), as it would chronologically follow referral.

9.2.3 *Screening*

During the screening stage, EFSA would be charged with identifying the most appropriate assessment approach(es) under which to gather knowledge on the relative threats (and, if deemed necessary, relative benefits) of Cry1Ab maize. Screening applies governance principles such as openness and effectiveness, as well as precaution, in order to characterise key features of different threats so as to determine the most effective assessment approach(es). Stakeholders are involved in screening through the review process that sets the detailed criteria against which the threats are screened. The first set of criteria gauges whether the threat is certainly and unambiguously serious, and therefore requires a presumption of prevention. As outlined in Sect. 4.2, the criteria include carcinogenicity, mutagenicity and reprotoxicity in food components or residues (as already embodied in existing regulatory initiatives in this field, such as the 2001 European Commission's Chemicals White Paper, CEC 2001b). Beyond this, attention might extend to further health threat criteria such as endocrine disruption, neurotoxicity, asthmagenicity or sensitising potential.

Screening for certainly and unambiguously serious threats in the Bt maize case study would therefore proceed according to the following criteria:

(a) Is there clear evidence of carcinogenicity, mutagenicity, reprotoxicity in components/ residues?
(b) Is there clear evidence of virulent pathogens?
(c) Is the new food associated with any violation of any risk-based concentration thresholds and/or legal standards?

Box 1. Possible outcomes for Cry1Ab maize case study – certainly and unambiguously serious threats?

The case study could proceed in the following way:

- Clear evidence of carcinogenicity, mutagenicity, reprotoxicity in components/ residues?
  - Based on current scientific understanding, the answer to this criterion is likely to be negative.
- Clear evidence of virulent pathogens?
  - Based on current scientific understanding, the answer to this criterion is likely to be negative.
- Violation of risk-based concentration thresholds and legal standards?
  - Based on current scientific understanding, the answer to this criterion is likely to be negative for a new Cry1Ab maize similar to Bt11 or Mon810, however the criterion could be triggered for Bt10 contamination or for another Bt maize event that included an antibiotic resistance marker gene previously assessed to have the potential to cause adverse effects on human health or the environment (see below).

Possible outcomes for these criteria are outlined in Box 1. As with all of the illustrative boxes in this chapter, these are merely suggestions by the author; EFSA might respond differently if confronted with the same criteria. If, under an application under Regulation 1829/2003, EFSA delivered a negative answer to all of these three criteria for the Cry1Ab Bt maize products being put forward, the screening process would bypass the option of presumption of prevention and proceed to the next set of criteria that gauge the scientific uncertainty surrounding the product.

A positive response might only potentially be expected to the third criterion, if, like Bt10, the Cry1Ab maize in question contained the ampicillin resistance marker *bla* (encoding beta-lactamase), which has been included in Group II by the EFSA GMO Panel (i.e. should be restricted to field trial purposes and should not be present in GM plants to be placed on the market) (EFSA 2004b), and should have been phased out by 31 December 2004 according to Directive 2001/18/EC. As such a Bt10-like incident would require a presumption of prevention, and, rather than passing on to further assessment, would (in the absence of mitigating factors) demand the immediate prevention of any placing on the market. A presumption of prevention in this case could be enacted in combination with a programme of testing for contamination and restricting imports at ports of entry. Potential mitigating factors to checking maize imports at sites of entry include cost considerations and the possibility of provoking trade disputes. There might also be indirect economic implications from the delays in food imports reaching processors and retailers.

If a presumption of prevention were not followed, the process would move on to screen for uncertainty. Screening of the threats of Bt maize using the criteria for uncertainty would be derived from the following:

(a) Are there scientifically founded questions concerning the status of the theoretical foundations of the disciplines bearing on the characterisation of the threat?
(b) Are there features of the food or food component in question which are substantively novel, in the sense that they involve characteristics or properties that are in some sense unprecedented?
(c) Are there scientifically founded questions concerning the completeness or sufficiency of the particular scientific models bearing on the characterisation of the threat?
(d) Are there scientifically founded questions concerning the applicability to the context in question of the particular scientific models used to characterise the threat?
(e) Are there scientifically founded questions concerning the applicability to the context in question of the data sets bearing on the characterisation of the threat?
(f) Are there scientifically founded questions concerning the quality of the data sets bearing on the characterisation of the threat of a kind that is not susceptible to probabilistic treatment?
(g) Do there exist any indirect, interactive or synergistic causal mechanisms of a kind that may not fully and confidently be characterised by probabilistic techniques?

Possible outcomes of such a screening are outlined in Box 2. If EFSA were to judge the answers to any of the above questions as likely to be positive, an approach of precautionary assessment would be triggered (see below in the section on assessment). Whether or not the outcome of the above screening criteria for uncertainty was positive, the product would next proceed to be screened for criteria for socio-political ambiguity.

Box 2 Possible outcomes for Cry1Ab maize case study – uncertainty?

The case study could proceed in the following way:

- Scientifically founded doubts on theory?
  - Based on recent scientific debates, the answer to this criterion might be positive – although most regulatory scientists disagree, there is a small minority of scientists who believe our knowledge of certain effects of plant transformation on food safety is still limited (e.g. as a result of transformation-induced mutations (see Latham, Wilson, & Steinbrecher 2006)).
- Novel/unprecedented features of the food?
  - Based on recent scientific debates, the answer to this criterion might be positive – the process by which the GM maize has been produced is relatively novel, and some of the components (e.g. truncated Bt toxin, marker gene products) have not previously made up significant components of the human diet (although the relevance of these facts is disputed by many scientists).
- Scientific doubts on model sufficiency or applicability?
  - Based on recent scientific debates, the answer to this criterion might be positive – although most regulatory scientists are content to adopt an approach that focuses on substantial equivalence, proximate analysis, acute toxicity tests and QSAR (quantitative structure–activity relationships), other experts have suggested that, due to the uncertainties surrounding the process of genetic modification, we require more specific tests on the range of metabolites produced in the plant when grown under various environmental conditions and immunological tests to assess the allergenicity of such metabolites (see Spök et al. 2003).
- Scientific doubts on data quality or applicability?
  - Based on recent scientific debates, the answer to this criterion might be positive – although most regulatory scientists disagree, some scientists believe that toxicity tests should be longer than the acute mouse studies cited in most GM maize dossiers, and should be carried out using the Bt maize itself rather than the Bt toxin produced in E coli or other GM plants as in some dossiers (see e.g. Freese & Schubert 2004).
- Indirect, interactive, or synergistic causal mechanisms of a kind that may not fully and confidently be characterised by probabilistic techniques?
  - Based on recent scientific debates, the answer to this criterion might be positive – although most regulatory scientists disagree, there is a small minority of scientists who believe that there could be possible indirect impacts on human health from long-term consumption and use of certain GMOs. Concerns relate to a number of unanswered (and unasked) questions including those highlighted above (Traavik & Heinemann 2007).

Screening criteria for socio-political ambiguity would, during the review process, be developed on the basis of the following:

(a) At the level of individual constituencies, is there a perceived threat of harm on a catastrophic scale (individual criterion)?
(b) Where there is disagreement between regulatory agencies and/or Member States, are there aspects of these institutional conflicts ostensibly unrelated to scientific uncertainty (institutional criterion)?
(c) With regard to the news media, are there signs that the threat in question is subject to a pronounced degree of amplification (amplification criterion)?
(d) At the level of society as a whole, are there signs of adverse effects in terms of social justice in the distribution of threat or in terms of manifest political mobilisation on the part of particular public constituencies (social criterion)?

Box 3 Possible outcomes for Cry1Ab maize case study – ambiguity?

- Divergent individual perceptions of risk?
  - The answer to this criterion is certainly positive. Formal studies of conflicting risk perceptions are available (e.g. see PABE Project – Public Perceptions of Agricultural Biotechnology in Europe, as well as complex opinions described in the various special Eurobarometer surveys on biotechnology), and these divergences would in addition be obvious based on discussions/input of stakeholders and the diverse perceptions of risks adopted by certain food safety organisations and industry groups.
- Institutional conflict between different administrative agencies?
  - The answer to this criterion is most likely to be positive. Although most administrative agencies in the EU have approved Cry1Ab maize as safe for human consumption, there have been advisory Competent Authorities that assessed the risks from Bt176, Mon810 and Bt11 to be unacceptable (e.g. Austria) and some disagreements at national level (e.g. between the French Commission for Genetic Engineering, CGB, and the French Food Safety Agency, AFSSA, in French case over Bt176).
- Amplification effects in news media?
  - The answer to this criterion is most likely to be positive (for GM food issues in general), although it is unclear whether there exists evidence to prove this, especially as far as Cry1Ab maize is concerned. One of the products has appeared (in its own right) in a negative light in the non-specialist press (Le Monde, 14th May 2004).
- Social justice concerns, distributional issues or political mobilisation?
  - There have been no specific concerns for low-income families, or particular groups (e.g. vegans); however, the answer to this criterion is undoubtedly positive. Although there may not have been mobilisation against Cry1Ab maize specifically, there has clearly been political mobilisation around the GM issue, and this would be obvious from stakeholder consultation.

Any positive responses from EFSA to these screening criteria would illustrate the need for concern assessment (see Chap. 4 on assessment for details). If the answers to screening criteria for socio-political ambiguity were all negative, concern assessment would not be judged appropriate or necessary. If neither uncertainty nor socio-political ambiguity criteria were triggered, conventional risk assessment would be the appropriate means by which to gather knowledge regarding the threats posed by the Cry1Ab Bt maize product.

9.2.4 *Terms of Reference*

At this point, based on the outputs of the screening process carried out by EFSA, the Interface Committee draws up specific and detailed terms of reference outlining the form of assessment needed in order to inform decision-making.

Terms of reference might specify:

- The precise scientific questions demanded of EFSA with respect to each threat identified, including the type of research that would be necessary and sufficient to address any uncertainties raised in the screening process, whether it be in vitro tests including quantitative modelling, in vivo screening, QSAR/molecular modelling, human (phase I/II/III pharmaceutical-style) tests/laboratory-based animal toxicity tests, as well as the forms of extended risk assessment to be employed subsequently in the characterisation of hazards and exposures, in order to calculate probabilities and magnitudes (and, therefore, risk).
- The primary issues of concern as identified by individual, institutional, amplification and social criteria (see above) of socio-political ambiguity, and the preferred methods for investigating these during the assessment phase. This might include focus groups, surveys or analytic-deliberative techniques.

The governance framework as it is advocated in this book envisions that the terms of reference are set jointly by the Commission and EFSA in cooperation with key stakeholders (through the Interface Committee). Furthermore, the proposed framework would see the draft terms of reference displayed in the Internet Forum in order to provide affected and interested actors with the opportunity to provide input (for specific details about the tasks and structure of the Interface Committee and the Internet Forum, see Chap. 6). This is especially important in cases where there is uncertainty and/or ambiguity. The institution formally referring the case to EFSA should be involved in producing the final document and, in order to fulfil the principle of accountability, should be able to provide a justification for specific changes between the draft terms of reference and the final version sent to EFSA.

9.3 Assessment

Based on the screening and terms of reference above, EFSA is then charged with carrying out the process of assessment, which gathers the relevant knowledge to feed into evaluation and management. Under current conditions, firms aiming to place a new type of genetically-modified Bt maize on the market would be required to provide a specified set of data in support of their application, which are then reviewed by the relevant EFSA scientific committee(s). Under the General Framework put forward in this book, such data are further specified in the terms of reference developed by the Interface Committee, and would be supplemented by EFSA by drawing on the peer-reviewed literature. Where necessary, EFSA should be able to further commission–external institutions to carry out investigations deemed necessary to address the terms of reference associated with a particular threat or threats.

In order to demonstrate the innovative forms of assessment proposed in this framework, we will assume that the two approaches to assessment to be followed in the case of Bt maize might be precautionary assessment and concern assessment. *Precautionary assessment* would follow from the referral process, which engaged multiple stakeholders in order to frame the questions asked and the information sought in assessment. Stakeholders such as consumers at large, consumer rights organisations, community groups, representatives of different geographic regions, public health agencies, including food research institutes, regulatory agencies such as the different national food authorities, trade associations, business, e.g. food industry, farmers, retailers, labour unions, environmental advocacy organisations, religious groups, educational and research institutions would have had the opportunity to contribute their respective knowledge to the process, minimising the likelihood of institutional ignorance. The input of these actors could be employed at the later stage of evaluation, in the interpretation of assessment outputs.

In order to make the precautionary assessment effective and efficient, it should be organised not necessarily at the level of the whole product but at the level of the scientific issues around which uncertainty remains. These might be the genetic modification process itself, the use of a particular promoter (e.g. CaMV35S) that drives the expression of the transgene products, or allergenicity potential resulting from the presence of a novel protein – the truncated Bt toxin –in the maize. Other assessment activities will necessitate the use of the whole product under various alternative conditions. For example, the scope of assessment could be extended to include a range of additive, cumulative and synergistic effects, addressing mixtures, derivatives, and reaction products (probabilistic approaches to the assessment of multiple toxins have been investigated by researchers from subproject 3 of the SAFE FOODS project (van der Voet, de Mul, & van Klaveren 2007). Alternatively, assessment might investigate the impact of the proteomic and metabolomic make up of the genetically-modified crop under various agricultural management conditions (the impact of agricultural regimes on protein profiles of potatoes has been investigated by subproject 1 of the SAFE FOODS project (Lehesranta et al. 2007).

A number of additional provisions can be used to investigate these uncertainties, and to directly address the more intractable forms of societal ignorance. Institutional trends and compliance issues should be systematically investigated in order to examine the assumptions underlying various policy options. The explicit examination of both the pros as well as the cons associated with the products or technologies presenting the threats in question, including consideration of technical substitutions, distributional issues around specific threats and benefits, is also required. Related to this, precautionary assessment should include the detailed and balanced comparison of contending merits and drawbacks of any design or policy options that present functional alternatives to the product or technology in question (including inaction and the *status quo*) and consideration of risk–risk trade-offs and, if there are any, better ways to provide the goods or services in question.

Reflecting a general principle of precaution, this form of assessment should adopt a conscious shift in the burden of persuasion, such that it is those who wish to implement the technology or product in question who must resource the acquisition of relevant data and sustain an argument as to the acceptable nature of the associated threat, subject to an appropriate level of proof. Finally, precautionary assessment should adopt an explicit focus on the extent to which the policies, technologies or products under scrutiny display properties such as flexibility, adaptability, reversibility, and diversity (all of which offer different ways of hedging against exposure to any residual societal ignorance that has not been addressed by the other elements in precautionary appraisal.)

At a point when all identified uncertainties have been clarified to the chosen level of safety, a process of extended risk assessment can be employed to reach probabilistic assessments of risk from the various options.

The discursive *concern assessment* process following the identification of ambiguity around Bt maize should aim to explore varying beliefs and values as they affect lifestyle choices, visions of the future and thus the decisions to be taken. The overall aim of such processes is to find solutions that cause least infringement of any group's values and beliefs and to clarify collectively valued benefits of the course of action.

It is important to consider the specific threats associated with the product that have triggered the criteria for ambiguity – i.e. is there ambiguity at the level of the specific product (Cry1Ab maize in human food), around specific options for its introduction (e.g. unlabelled, labelled), around the product's use elsewhere in the food chain (in animal feed or for cultivation in general), or around the product as part of a whole class of new technologies (GM food in general)? In this particular case ambiguities arise at all these levels, a finding that goes on to influence the level at which concern assessment should take place. In some cases the most efficient scale at which to consider the product would be through aggregating it with other products to the level of GM food (or at least GM maize) in general. Ideally, this aggregated form of concern assessment (as with the precautionary assessment above) would be carried out the first time a new type of food/technology (GM food/ Bt maize variety) was referred to EFSA. Screening would be able to identify the novel characteristics of the food/technology and inform the appropriate level for 'bundling' (cp. Chap. 6) foods/technologies together. In this particular case, we have seen that concern assessment could be carried out at the level of GM food in general, and thus, although it is too late to conduct an initial concern assessment for GM food in the current context, the General Framework would have advocated such a concern assessment at the time of the first applications for genetically-modified

Cry1Ab Bt maize (and other genetically-modified foods) under Directive 1990/220/EC in 1994.

Based on the issues raised during screening and the specific terms of reference, the concern assessment process might need to address issues such as:

- Concerns over whether the product would be introduced as processed food (including different methods of processing) or as unprocessed food (e.g. sweetcorn)
- How consumers would view various labelling conditions under which the product could be introduced (including different allowable thresholds of 'contamination'): positive – 'this product contains...' or negative – 'this product does not contain...'), forms of labelling, the level of detail ('this product may contain GM materials/Bt maize/Cry1Ab maize/specific events/maize modified with genes ... from ...'), legal arrangements for labelling and traceability (voluntary, enforcement, liability and redress)
- The potential for any economic impact of the approval of the product on various sectors of the agricultural and food sectors
- The potential impact on export markets or trade partners of various conditions under which the marketing of the product could be approved or prevented (e.g. impacts on exporters in the developing world whose economic well-being might depend upon access to EU markets)
- Levels of awareness and associated concerns around the institutions charged with enforcing and monitoring the safe use of the product, post-approval, and levels of trust in these institutions[4]
- Public positions on possible ethical and moral objections to (or reasons to support) genetic modification in general

Under the assumption that criteria for uncertainty and ambiguity were triggered in this case, both precautionary and concern assessment would be carried out. The precautionary assessment and concern assessment described above would lead directly to an extended risk assessment at a time when uncertainties would have been clarified to a point determined by the chosen level of safety (as specified by the terms of reference). As stated in earlier chapters, this risk assessment should be comprehensive, detailed, systematic, and rigorous, and might involve not only the scientific approaches listed above, but could also include probabilistic techniques such as stochastic, Monte Carlo, Bayesian and/or exposure modelling. When using these techniques – which are usually associated with conventional risk assessment – assumptions, experimental designs and findings of the risk assessment (along with sensitivity analyses) should be transparently communicated to allow for peer review not only by the scientific community but by wider groups of experts from interested stakeholders. As precautionary assessment would employ not only the broad techniques described above but also these probabilistic techniques associated

[4]Similar issues to this have been studied by researchers from subproject 4 of the SAFE FOODS project. See van Kleef et al. (2006).

with extended risk assessment, there would be no further requirement for an additional stage of conventional risk assessment.

9.4 Evaluation and Management

The process of evaluation uses the outputs of these various forms of assessment to judge the tolerability or acceptability of a given threat and, if deemed necessary, to initiate the appropriate management process. The process is carried out (in an advisory function) by the Interface Committee, and supported by the deliberations on the Internet Forum (the Internet Forum being managed by the Commission). Based on assessments of the likely consequences for human health or other relevant endpoints and the concerns that individuals, groups or different cultures may ascribe to a given food safety problem, the stakeholders represented on the Interface Committee will bring a range of values to bear on the decision over whether the threat is intolerable (i.e. the food or technology assessed needs to be abandoned or replaced), tolerable (i.e. management measures need to be formulated in order to reduce or handle the threats in question), or acceptable (i.e. management measures to reduce the threat in question are deemed unnecessary). The deliberations should include evaluation not only of pros and cons for human health (as assessed through precautionary assessment or risk assessment) but also wider social–political factors (including labelling and traceability conditions, economic considerations, perceptions on the distribution of risks and benefits and how these are managed, social mobilisation and conflict potential, as well as moral and ethical considerations) as assessed through the concern assessment detailed above.

If the conclusion is that management measures are required, the Interface Committee will come up with a recommendation for the most appropriate management approach from the four approaches – prevention, precaution, risk and concern – outlined in Chap. 6. It is important to note here that the approaches to assessment described above do not automatically determine the management approach followed thereafter – the management approach is chosen only at the stage of management in consideration of the advice provided by the Interface Committee at the evaluation stage. Furthermore, the management approaches do not automatically determine the management measure to be selected, however point towards certain ones that are likely to be appropriate (for the list of management approaches and associated measures see Table 5.2). If the evaluation process identifies areas of salient knowledge not covered in the assessment process, there would exist the potential to feed back into new terms of reference.

In order to further guard against surprises arising as a result of societal ignorance, selection of management options should provide flexibility and reversibility in the event that unexpected negative impacts arise at a later date. In the case of long-term GM food safety, this might include a labelling and traceability system, a surveillance system for unanticipated impacts, maintenance of non-GM supply chains, diversity within the food production system, or other features that build resilience.

In the management process itself, as described in Sect. 5.3, possible management measures would be identified, assessed, evaluated, and selected. Following the selection of the appropriate measures, they are implemented. The monitoring of how these measures perform in practice represents a final and continuous stage, potentially feeding back into the governance cycle through the process of review.

For the Cry1Ab Bt maize case, the appropriate management measures (drawn from those outlined in Chap. 5) might consist of:

- Approval or otherwise of the placing on the market of the product, under one or more of the following (or other) conditions
- Technical standards and limits that prescribe the permissible threshold of concentrations of the product (which may merely apply those thresholds specified in Regulation 1830/2003)
- Performance standards e.g. for processes of labelling and traceability
- Insurance and liability arrangements
- Close monitoring of adverse effects
- Selecting functional equivalents which, under certain conditions, present less risk or uncertainty

These would then be assessed on the basis of their effectiveness, efficiency, minimisation of external side effects, fairness, sustainability, political and legal implementability, ethical acceptability, and public acceptance. In the case of Bt maize, if authoritative research were to show that the marketing of genetically modified maize was preferred/acceptable (as has been claimed by a study in Canada (Powell, Blaine, Morris, & Wilson 2003), political implementability and public acceptance would not pose a barrier to approval. If, on the other hand, studies showed a significant proportion of the population to be generally opposed to the introduction of such foods (Gaskell et al. 2006), this information would need to be included in the evaluation phase and be weighed against other potentially positive evidence on physical or economic risks and benefits. A high degree of public concern can be reason enough to justify a ban, if the other arguments do not outweigh this negative impact, although such a decision would have obvious implications at the WTO. Recent work has suggested that 'there remains considerable scope for greater recognition within SPS jurisprudence of the significance of public opinion in decision-making about risks to human health and environment, in a way that combines scientific and non-scientific aspects of decision-making about risk' (Foster 2008).

The outputs of these assessments (usually carried out by experts and stakeholders) would be evaluated (i.e. weighted in their importance) by politically legitimate and accountable decision-makers, who would then be responsible for selecting the most appropriate management measures. The reasons and justification (including assessment and evaluation outcomes) for this particular selection of management measure(s) should be posted on the Internet Forum to promote the principles of accountability, coherence, and consistency. Following selection of the appropriate measures, these would then be implemented by the responsible institutions at the EU, Member State or other administrative level. These institutions could also be responsible for monitoring, with the aid and support of a wider group of actors including the corporate sector and various concerned non-governmental organisations.

As mentioned above and at the beginning of this case study, the General Framework outlined here is designed so as to be flexible and responsive to emerging scientific information and changing socio-political conditions. The importance of monitoring these factors, and the possibility that assessments may be re-framed as a result (through the process of review) cannot be underestimated, especially in an area such as genetically modified foods, which is subject to rapidly evolving science and technology (both in production and regulation) and subject to intense socio-political controversy.

Chapter 10
Summary: Key Features of the General Framework

M. Dreyer, O. Renn, A. Ely, A. Stirling, E. Vos, and F. Wendler

European food safety governance is an evolving system. With the advent of strong pressures and many associated recent reforms, new challenges are posed for implementation. Demanding questions are raised over the means to achieve more practical and effective cooperation between diverse actors in regulatory decision-making processes. Our own research into this field began by observing the special characteristics of food regulation and risk governance and the dynamic nature of the changes which these have undergone in Europe since the mid 1990s. Shaken by a series of safety scares and controversies, the European food governance system is undergoing a far-reaching process of review and reform – which already entails powerful imperatives for greater stakeholder involvement and more deliberate attention to uncertainty. In the hope of contributing constructively to this process, the present *General Framework for the Precautionary and Inclusive Governance of Food Safety* is intended to offer a thorough analysis of the currently emerging regulatory changes. This in turn provides a basis for a detailed and carefully measured set of suggestions towards achieving greater coherence, direction, and purpose in current arrangements for food safety governance.

The main argument underlying this approach is readily stated. In order to improve both the effectiveness and the democratic legitimacy of food safety governance in Europe, it is necessary to distinguish more carefully between different aspects and contexts. In particular, strongly contrasting implications are presented by food safety threats that may be seen (respectively) as: "routine" in nature, definitely "prohibitive" in their consequences, or in some way "intractable" – either because their implications are scientifically uncertain or are socio-politically ambiguous. Each of these broad aspects and contexts of food safety demands different kinds of attention and different modes of co-ordination between specialist experts, political decision-makers and corporate and civil society actors. Each in turn therefore also requires at least partly distinct technical methodologies, deliberative processes and institutional configurations. The main contribution of the present work has been to scope out a minimal and straightforward way in which these complex demands might be reconciled.

In undertaking this task, we have paid special attention to the compatibility of the proposed procedural and institutional reforms with the current EU legal and policy framework. The purpose of the present chapter is to clarify the way in which

M. Dreyer and O. Renn (eds.), *Food Safety Governance*,
DOI: 10.1007/978-3-540-69309-3_11, © Springer-Verlag Berlin Heidelberg 2009

we have pursued this objective. It begins with a summary of the *key features* of the proposed General Framework and highlights the way in which these features relate to established principles of food safety governance as enshrined in the General Food Law and high profile general agendas around the governance of European Union institutions.

The General Framework builds around a logical structure founded on four consecutive stages, which we call *framing*, *assessment*, *evaluation*, and *management*. Different forms of cross-cutting activities of participation and communication variously accompany each of the four stages. A crucial point at the outset, is that this four-stage design reproduces the *separation* of assessment and management activities as specified in the General Food Law. For our part, despite the many interactions and interdependencies, we are also persuaded of the merits of establishing a clear conceptual and functional distinction between assessment and management. Assessment and management involve "different goals, kinds of expertness, and operating principles" (NRC 1983: 151). But – again in common with other prominent established approaches to risk governance – this General Framework also adds two further stages to the process: *framing* and *evaluation*. These two stages constitute *mediating activities* between processes of assessment (focused on knowledge generation, collection and interpretation) and management (focussing on value-laden decision-making in a jigsaw puzzle of facts, uncertainties, stakeholder interests, and public concerns). Framing provides guidance concerning the articulation of the "problem" to be addressed, the boundaries of the investigations to be conducted and procedures necessary for further handling of the food safety threat in question – especially during assessment. It is during framing, for instance, that the terms of reference are specified for assessment. This task needs to be governed by societal values (stating the goals, objectives, and contextual conditions) and inspired by what we already know about the threat (suspected impacts, exposure, persistence, and others).

Similarly, it is during evaluation, that judgements are developed concerning the tolerability or acceptability of the threats under scrutiny. This necessarily follows the elicitation and definition during assessment, of firmer understandings of salient evidence, residual uncertainties, relevant ambiguities and persistent ignorance. It also requires the articulation of judgemental competence for making the necessary trade-offs between threats, benefits and other relevant impact aspects, as viewed under multiple perspectives. It is in these ways, that framing and evaluation may be seen as distinct *hybrid activities*, in the sense that they draw on both political and socio-economic considerations as well as scientific knowledge. Whether or not this is explicitly acknowledged, knowledge and values are closely intertwined in these activities. It is for this reason that many stakeholders (for different reasons) identify the need for improved interaction between assessors and managers. Such views feature prominently and repeatedly as part of general food safety governance debates, and are strongly represented in the deliberative processes undertaken as part of the development of the present proposed framework. Under current regulatory arrangements, these hybrid activities are typically carried out in a manner which lacks transparency, with the exercise of evaluation and practice of framing being largely implicit and ad hoc and the associated responsibilities correspondingly unclear and under-accountable. By highlighting the distinct character of framing and evaluation, the present framework does not therefore introduce new activities, but rather clarifies and renders more explicit the terms for cooperative interaction between managers, assessors, and key stakeholders. In order to remain operational in existing institutional contexts, this more detailed attention and transparency is achieved under the auspices of a single new dedicated institutional innovation. We recommend that this takes the form of what we call an *Interface Advisory Committee* (cp. Sect. 6.4.2), but a variety of options are discussed.

The four-stage design proposed in this General Framework thus retains the basic form of familiar institutional activities, but avoids associated necessity for naïve assertions over the separation of facts and values in assessment and management. However, by retaining a respect for the distinct forms of attention required in generating knowledge and eliciting values, this approach also avoids concerns over "post-modern" or "relativist" views, under which such activities are regarded as homogenous. Crucially (given prominent wider aspirations in European governance), it is by distinguishing these different activities that the Framework provides for greater *accountability* in the ways in which knowledge and value inputs are articulated in management decisions. This formalisation of framing and evaluation stages in the food safety governance process also improves accountability over the allocation of responsibilities for essential governance activities.

A further important element that flows from this, is the development of the presently largely tacit role of *screening*. This procedure is designed to tailor the overall governance process to address key attributes of food safety threats in the most appropriate, effective and efficient way. It is by this means that the proposed framework may therefore claim to enhance the governance process under the European Commission's principle of *coherence* in governance. More specifically, the screening procedure operates by applying systematic criteria to allow *distinctions between attributes of seriousness, uncertainty, and ambiguity*. These distinctions then guide the selection of the appropriate approach(es) to assessment and also assist in decision-making over the extent and form of any extended participation that may be required. It is through application of this screening procedure that the Framework determines the conditions under which the "more comprehensive risk assessment" referred to in the General Food Law may be required, whilst clarifying the relationship with other forms of assessment, (including quantitative risk assessment and concern assessment). These conditions of scientific uncertainty (addressed in precautionary assessment), and socio-political ambiguity (addressed in concern assessment), are the key ingredients for intense and persistent conflicts between regulators and corporate and civil society actors over new and emerging food production technologies. They are of special relevance in contemporary societies characterized by plural world views and multiple firmly held beliefs about what constitutes "good" food as well as by plural knowledge claims. The precaution-based governance approach and the concern-oriented governance approach are designed to address these issues with *greater analytical rigour* and *more effective deliberation* – and thereby offer means to reduce the intensity and/or persistence of food safety conflicts.

Under a second European Commission governance principle, concerning *consistency*, the four-track design of this framework (*risk-*, *precaution-*, *concern-* and *prevention-based approaches*) promises a more effective means to achieve this quality, than does an undifferentiated assessment process. It does this by providing for more explicit and deliberately focused attention to crucial differences between different food safety threats and contexts (e.g.: extending beyond morbidity and mortality). In the absence of this attention, it becomes more difficult to demonstrate consistency in the handling of *multiple challenges* in food safety governance. With the concern assessment approach, which is, for instance, triggered under the condition that criteria of socio-political ambiguity apply, the General Framework expands the set of criteria for assessing, evaluating and managing food safety threats that have dominated conventional concepts of food safety governance. Public values, social concerns, and perceptions of food safety issues are included in the governance process. It is becoming increasingly well recognised that these issues are often just as important as expected mortality and morbidity frequencies for identifying, understanding, and managing food safety threats. Clearly, the social scientific analysis involved in concern assessment should be submitted to the same rigour of methodological scrutiny and peer review as any other scientific activity.

It is in delivering the crucial quality of "consistency", that the proposed initial step of screening also comes to the fore. However, screening also invokes two further European Commission governance principles, by contributing to the *timeliness* of food safety management and the *effectiveness* of the procedures applied. Screening addresses these criteria by means of more explicit and systematic pursuit of an activity that is already quite well recognised as "preliminary assessment". With assessment itself differentiated as discussed above, screening provides a means to identify the most appropriate approach to a more detailed assessment and to help prioritise attention to diverse threats. This is achieved by applying criteria of seriousness, uncertainty, and ambiguity in order to inform the detailed terms of reference.

A recent publication on the role of expert advice in the governance of science and technology states rightly that "public engagement is not a stage of governance that can be completed, tidied up and filed away" (Stilgoe et al. 2006: 53). This raises the highly exigent question of how to incorporate the perspectives and specialized knowledge of interested and affected parties into the governance process. The food safety interface institutions and screening procedures envisaged in this General Framework offer a more systematic approach to fully honouring this imperative – one that is also recognised in a European Commission governance principle of *participation*. It does this by providing both for permanent general processes and more flexible participatory mechanisms. These are incorporated in such ways as to ensure better coordination of assessment and management, and more effectively and transparently to address the concerns of corporate and civil society actors throughout the governance process. In particular, the interface institutions present standing platforms for deliberation over major elements of the governance process, with the *Internet Forum* being the most inclusive in terms of both the governance elements which it opens up for deliberation, and the voices which it invites for engaging in this deliberation. The Interface Committee is recommended as an additional liaison institution, which specifically accounts for ways that framing and evaluation activities cut across assessment and management. The selection of a few "key stakeholders" to sit on the Interface Committee will inevitably provoke questions over representativeness, power, accountability and fairness (see Chaps. 11 and 12 for more detail). However, the existence of the second interface institution – the Internet Forum, which is more inclusive – might alleviate such concerns. A further remedy in this regard, is the provision under this Framework that threats associated with high levels of uncertainty and/or ambiguity receive more detailed, intensive and specifically tailored participatory procedures. It is by striking this balance between permanent general levels and more targeted specific forms of participation, that this Framework is designed to be practical and operational in the face of the many hundreds of requests typically received by European institutions for opinion in one single year.

The inclusion of these participatory elements in the design of the present General Framework can also help to achieve a broader and more structured engagement between the EFSA and the Commission. By including a diversity of interests in these interface activities, greater, more transparent and systematically structured attention is afforded to the relevant knowledges, interests, values and preferences. The governance process as a whole is thus rendered more sensitive and responsive, whilst at the same time avoiding overburdening through excessive and exhaustive levels of participation on every food safety issue. The Internet Forum, in particular, thus helps incorporate principles of *transparency* (from the point of view of third parties) and so respect the European Commission's governance principle of openness. By bringing communication with affected and interested groups at the assessment/management interface into the more formal domain, stakeholder representation in the Interface Committee would render engagement with stakeholders more transparent and symmetrical. Together with the recently established major stakeholder fora of the European Commission and EFSA[1], the Interface Committee could act as a counterbalance to more informal and bilateral channels for lobbying and the peddling of influence by powerful interests. As such, stakeholder representation in the Interface Committee, in combination with the web-based consultations and deliberations through the Internet Forum, might enable greater *accountability* over the ways in which decision-making relates to potentially contending positions in food safety debates.[2]

The Internet Forum and the Interface Committee are thus at the core of our suggestions for institutional reform. They are designed to work as an innovative food safety interface structure which can improve the politics–science–society coordination throughout the governance process. A further modest institutional innovation to improve capacities to conduct the tasks of screening and concern assessment, is the

[1] These are the European Commission's Advisory Group on the Food Chain and Animal and Plant Health and EFSA's Stakeholder Consultative Platform.

[2] Hence, our argument is that stakeholder representation in the Interface Committee would not enhance but reduce the risk of regulatory capture by industry interests; for a different view see Gabbi (2007) and Alemanno (2008).

proposal of a *Screening Unit* within EFSA. This would work as a clearing-house between the secretariat of EFSA and the various scientific panels at the screening stage, and the establishment of a new EFSA *Panel on Concern Assessment* tasked with providing specific expertise to address questions of socio-political ambiguity in assessment. These four new minimal institutional reforms are deemed essential for facilitating the working of the proposed procedural reforms. We have made an effort to keep the number of institutional innovations to a minimum and to design them in such a way that they can be easily integrated into the current governance structure. The Internet Forum, which is our basic recommendation for establishing a food safety interface structure, links to the increasing use of the Internet for documentation and for eliciting stakeholder feedback by EFSA and the Commission. While the Internet Forum could also be set up as the sole interface institution, we would recommend establishing, in addition, the Interface Committee. This in turn might be in form of an Advisory Committee (which is our preferred variant), or a Steering Committee (which would provide framing and evaluation activities with a formal footing). The EFSA Screening Unit would not address the screening questions itself, but would pass on requests for screening to the different scientific units and could thus be easily integrated into EFSA's current structure. The same applies to the Concern Assessment Panel. It would extend EFSA's scientific panels by one. It would take a fairly conventional format except that the Panel would comprise experts with a background in the social, psychological and economic sciences of a kind that would enhance EFSA's capacity for carrying out concern assessment.

Finally, it is important to note that the procedural innovations associated with this Framework include provisions to avoid overburdening the food safety governance system and overexploiting scarce financial and staff resources for making decisions. These provisions are the key to making the proposed framework *practical*. The proposed fourfold differentiated approach to assessment expands the assessment process only as judged appropriate under specified conditions (i.e.: in cases where scientific uncertainty warrants precautionary assessment or socio-political ambiguity warrants concern assessment). In other words, the Framework does *not* require that currently routine health risk assessment is expanded in every case by wholesale analysis of social, economic, and ethical impacts. This only occurs where the contexts of the particular cases in question are judged under screening to warrant this more expanded attention. Likewise, our recommended approach to participation implies extensive activity (beyond the Internet Forum and the Interface Committee) only under very particular conditions (namely those characterized by high levels of scientific uncertainty and socio-political ambiguity).

Of course, the bulk of the business handled under the proposed Framework would (as is the case at present) remain in routine quantitative assessment and conventional evaluation and management. The actors, procedures and institutions involved in undertaking these activities, will remain essentially the same under the proposed governance framework. For these routine kinds of case, no major additional efforts will be required to engage stakeholders, elicit concerns, examine uncertainties or develop sophisticated management measures. It is only under conditions where food safety threats are in any case already more complex and demanding (albeit without such systematic dedicated institutional or procedural provision) that the proposed General Framework envisages more elaborate activity. Given the experience of the past few years of European food safety governance, and the outcome of our own detailed analysis, the present research team is of the view that the more extensive, differentiated and sophisticated processes developed here offer the basis for a food safety governance system that is at the same time more scientifically rigorous, politically balanced and socially robust.

References (Part 1)

Ad hoc Commission "Neuordnung der Verfahren und Strukturen zur Risikobewertung und Standardsetzung im gesundheitlichen Umweltschutz der Bundesrepublik Deutschland" (2003). *Abschlussbericht über die Arbeiten der Risikokommission*. München: Bundesamt für Strahlenschutz.

Alemanno, A. (2001). Le principe de précaution en droit communautaire. Stratégie de gestion ou risque d'atteinte au marché intérieur. *Revue du droit de l'Union Européenne*, *4*, 917–940.

Alemanno A. (2008). The European food safety authority at five. *European Food and Feed Law Review 1,* 2–24, electronic copy available at: http://ssrn.com/abstract=1095703.

Amy, D. J. (1983). Environmental mediation: an alternative approach to policy stalemates. *Policy Sciences*, *15*, 345–365.

Ansell, C., & Vogel, D. (Eds.). (2006). *What's the Beef?: The contested governance of European food safety*. Cambridge: MIT.

Armour, A. (1995). The citizen's jury model of public participation. In O. Renn, T. Webler, & P. Wiedemann (Eds.), *Fairness and competence in citizen participation. evaluating new models for environmental discourse* (pp. 175–188). Dordrecht and Boston: Kluwer.

Atman, C. J., Bostrom, A., Fischhoff, B., & Morgan, M. G. (1994). Designing risk communication: completing and correcting mental models of hazardous processes (Part 1). *Risk Analysis*, *14*(5), 779–788.

Bacow, L. S., & Wheeler, M. (1984). *Environmental Dispute Resolution*. New York: Plenum.

Bandle, T. (2007). Tolerability of risk: the regulator's story. In F. Boulder, D. Slavin, R. Löfstedt (Eds.), *The tolerability of risk: a new framework for risk management* (pp. 93–104). London: Earthscan.

Baram, M. S. (1984). The right to know and the duty to disclose occupational hazard information. *American Journal of Public Health*, *74*, 385–390.

Barr, C. (1996). Fear not the art of risk communication. *Journal of Management in Engineering*, *12*(1), 18–22.

Beck, U. (1992). *Risk society toward a new modernity*. London: Sage.

Beck, U. (2000). The cosmopolitan perspective: sociology of the second age of modernity. *British Journal of Sociology*, *51*, 79–105.

Bennett, P. G., & Calman, K. C. (Eds.). (1999). *Risk communication and public health: policy, science and participation*. Oxford: Oxford University Press.

Bergström, C. F. (2005). *Comitology–delegation of powers in the european union and the committee system*. Oxford: Oxford University Press.

BfR (Bundesinstitut für Risikobewertung). (2005). *ERiC – Development of a multi-stage risk communication program*. In R. Hertel & G. Henseler (Eds.), BfR-Wissenschaft 02/2005. Berlin: Federal Institute for Risk Assessment (BfR).

Böschen, S., Dressel, K., Schneider, M., Viehöver, W., Wastian, M. (2005). *A review of institutional arrangements for food safety regulation in Germany. Food safety: administration of impossibilities or expression of modern and efficient governance?*, Deliverable 5.2.4 of

Subproject 5 of the EC Framework Programme 6 Integrated Project 'SAFE FOODS'. Munich: Süddeutsches Institut für empirische Sozialforschung (SINE).
Boholm, A. (1998). Comparative studies of risk perception: a review of twenty years of research. *Journal of Risk Research, 1*(2), 135–163.
Bostrom, A., Atman, C. J., Fischhoff, B., & Morgan, M. G. (1994). Evaluating risk communications: completing and correcting mental models of hazardous processes. *Risk Analysis, 14*(5), 789–798.
Breakwell, G. M. (2007). *The Psychology of Risk.* Cambridge: Cambridge University Press.
Bunting, C., Renn, O., Florin, M. -V., & Cantor, R. (2007). Introduction to the IRGC risk governance framework. *The John Liner Review, 21*(2), 7–26.
Burns, T. R., & Ueberhorst, R. (1988). *Creative democracy: systematic conflict resolution and policymaking in a world of high science and technology.* New York: Praeger.
Chess, C., Hance, B. J., & Sandman, P. M. (1989). *Planning dialogue with communities: a risk communication workbook, environmental communication research program.* New Brunswick, NJ: Rutgers University.
Christiansen, T., & Larsson, T. (2007). *The role of committees in the policy-process of the European Union.* Cheltenham: Edward Elgar.
Codex Alimentarius Commission (CAC). (2005). *Procedural Manual (fifteenth edition), Joint FAO/WHO Food Standards Programme.* Rome: World Health Organization/Food and Agriculture Organization of the United Nations.
Commission of the European Communities (CEC). (2000a). *Communication from the Commission on the Precautionary Principle*, COM (2000) 1, 2 February 2000. Brussels.
Commission of the European Communities (CEC). (2000b). *White Paper on Food Safety*, COM (1999) 719 final, 12 January 2000. Brussels.
Commission of the European Communities (CEC). (2001a). *European Governance. A White Paper*, COM 428 final, 25 July 2001. Brussels.
Commission of the European Communities (CEC). (2001b). *White Paper. Strategy for a future Chemicals Policy* (presented by the Commission), COM (2001) 88 final, 27.2.2001. Brussels.
Commission of the European Communities (CEC). (2003). Commission Regulation (EC) No 1304/2003 of 11 July 2003 on the procedure applied by the European Food Safety Authority to requests for scientific opinions referred to it. *Official Journal of the European Union*, L 185/6, 24.7.2003.
Commission of the European Communities (CEC). (2006). Commission Regulation (EC) No 575/2006 EC of 7 April 2006, amending Regulation (EC) No 178/2002 of the European Parliament and of the Council as regards the number and names of the permanent Scientific Panels of the European Food Safety Authority. *Official Journal of the European Communities*, 8.4.2006, L 100/3.
Corcelle, G. (2001). La perspective communautaire du principe de précaution. *Revue du marche commun et de l'Union Européenne, 450*, 447–454.
Covello, V. T. (1983). The perception of technological risks: a literature review. *Technological Forecasting and Social Change, 23*, 285–297.
Covello, V. T. (1992). Trust and credibility in risk communcation. *Health and Environmental Digest, 6*(1), 1–3.
Covello, V. T., & Sandman, P. M. (2001). Risk communication evolution and revolution. In A. Wolbarst (Ed.), *Solutions to an Environment in Peril* (pp. 164–178). Baltimore: Johns Hopkins University Press.
Covello, V. T., Slovic, P., & von Winterfeldt, D. (1986). Risk communication: a review of the literature. *Risk Abstracts, 3*(4), 172–182.
Covello, V. T., Sandman, P., & Slovic, P. (1988). *Risk communication, risk statistics and risk comparisons: a manual for plant managers.* Washington, DC: Chemical Manufactures Association.
Covello, V. T., McCallum, D. B., & Pavlova, M. (Eds.). (1989). *Effective risk communication: the role and responsibility of government and non-government organizations.* New York: Plenum.

Crosby, N., Kelly, J. M., & Schaefer, P. (1986). Citizen panels: a new approach to citizen participation. *Public Administration Review, 46*, 170–178.

DeFleur, M. L., & Ball-Rokeach, S. (1982). *Theories of mass communication*. New York: Longman.

Dehousse, R. (1994). Community competences: are there limits to growth? In R. Dehousse (Ed.), *Europe after Maastricht. An Ever Closer Union?* (pp. 124–125). Munich: Law Books in Europe.

De Jonge, J., van Kleef, E., Frewer, L. J., & Renn, O. (2007). Perception of risk, benefit and trust associated with consumer food choice. In L. J. Frewer, & H. van Trijp (Eds.), *Understanding consumers of food products* (pp. 534–557). Cambridge, UK: Woodhead.

De Marchi, B. (1995). Environmental problems, policy decisions and risk communication: what is the role for social sciences? *Science and Public Policy, 22*, 157–161.

De Sadeleer, N. (2001a). L'émergence du principe de précaution. *Journal des tribunaux, 6010*, 393–401.

De Sadeleer, N. (2001b). Le statut juridique du principe de précaution en droit communautaire: du slogan à la règle. *Cahiers de droit européen*, 79–120.

De Sadeleer, N. (2006). The precautionary principle in EC health and environmental law. *European Law Journal, 12*(2), 139–172.

DG SANCO (2005). Maximising the contribution of science to European health and safety, A DG SANCO Discussion Paper. Brussels, July 2005. http://europa.eu.int/comm/dgs/health_consumer/library/reports/science_health_safety_en.pdf. Accessed 20 August 2006.

DG SANCO (2007). General food law – principles. http://ec.europa.eu/food/food/foodlaw/principles/index_en.htm. Accessed 15 March 2007.

Dienel, P. C. (1989). Contributing to social decision methodology: citizen reports on technological projects. In C. Vlek, & G. Cvetkovich (Eds.), *Social decision methodology for technological projects* (pp. 133–151). Dordrecht: Kluwer.

Directive 2001/18/EC of the European Parliament and of the Council of 12 March 2001 on the deliberate release into the environment of genetically modified organisms and repealing Council Directive 90/220/EEC, *Official Journal of the European Communities*, 17.4.2001, L 106/1.

Douma, W. T. (2002). *The Precautionary Principle. Its Application in International, European and Dutch Law* (Doctoral dissertation, University of Groningen, 2002).

Dressel, K., Böschen, S., Schneider, M., Viehöver, W., Wastian, M., & Wendler, F. (2006). Food safety regulation in Germany. In E. Vos, F. Wendler (Eds.), *Food safety regulation in Europe: a comparative institutional analysis (Series Ius Commune)* (pp. 287–330). Antwerp: Intersentia.

Dreyer, M., Renn, O., Borkhart, K., & Ortleb, J. (2006). Institutional re-Arrangements in European food safety governance: a comparative perspective. In E. Vos, F. Wendler (Eds.), *Food safety regulation in Europe: a comparative institutional analysis (Series Ius Commune)* (pp. 9–64). Antwerp: Intersentia.

Dreyer, M., Renn, O., & Borkhart, K. (2006). *A summary report of a workshop with industry representatives, for the EC framework programme 6 Integrated Project 'SAFE FOODS', contribution to Subproject 5, 19 October 2006*. Stuttgart: DIALOGIK.

Dreyer, M., Renn, O., Ely, A., Stirling, A., Vos, E., & Wendler, F. (2007). *A general framework for the precautionary and inclusive governance of food safety, for the EC framework programme 6 Integrated Project 'SAFE FOODS', Interim Report of Subproject 5, 4 May 2007*. Stuttgart: DIALOGIK.

Dreyer, M., Renn, O., & Borkhart, K. (2007). *A summary report of a workshop with risk assessors (EU-Member State level), for the EC framework programme 6 Integrated Project 'SAFE FOODS', contribution to Subproject 5, 15 January 2007*. Stuttgart: DIALOGIK.

Drottz-Sjöberg, B. -M. (2003). *Current trends in risk communication theory and practice*. Oslo: Directory of Civil Defense and Emergency Planning.

EFSA (2004a). *Making risk assessment more transparent – information note to the advisory forum*, AF 08.04.2004.

EFSA (2004b). Opinion of the Scientific Panel on Genetically Modified Organisms on the use of antibiotic resistance genes as marker genes in genetically modified plants, Question No. EFSA-Q-2003-109. *The EFSA Journal, 48*, 1–18.

EFSA (2005). Opinion of the scientific panel on contaminants in the food chain on a request from the European parliament related to the safety assessment of wild and farmed fish, Question No. EFSA-Q-2004-22. *The EFSA Journal, 236*, 1–118.

EFSA (2006a). Transparency in risk assessment carried out by EFSA: guidance document on procedural aspects, prepared by a working group consisting of members of the Scientific Committee and various EFSA Departments, endorsed on 11 April 2006 by the Scientific Committee. *The EFSA Journal, 353*, 1–16.

EFSA (2006b). *Summary report of EFSA scientific colloquium 6: Risk-benefit-analysis of foods: methods and approaches*. Tabiano, Italy, 13–14 July 2006.

Ely, A., & Stirling, A. (2006). *A summary report of a workshop with NGO representatives, for the EC FRAMEWORK PROGRAMME 6 Integrated Project 'SAFE FOODS', contribution to Subproject 5, 2 November 2006*. Brighton: University of Sussex.

Environmental Protection Agency (EPA), Omen, G. S., Kessler, A. C., Anderson, N. T., et al. (1997). *Framework for environmental health risk management*, US Presidential/Congressional Commission on Risk Assessment and Risk Management, Final Report Vol. 1. Washington: EPA.

ESTO (2000). *On science and precaution in the management of technological risk*. In A. Stirling & V. Calenbuhr (Eds.), Final summary report of 'technological risk and the management of uncertainty' Project. Seville: European Scientific Technology Observatory (ESTO).

Faure, M. G., & Vos, E. (2003). *Juridische afbakening van het voorzorgsbeginsel: Mogelijkheden en grenzen, Gezondheidsraad*. The Hague: Dutch Health Council.

Fearn-Banks, K. (1996). *Crisis communications: a case book approach*. Mahwah, NJ: Lawrence Erlbaum.

Fiorino, D. J. (1990). Citizen participation and environmental risk: a survey of institutional mechanisms. *Science, Technology, and Human Values, 15*(2), 226–243.

Fischhoff, B. (1995). Risk perception and communication unplugged: twenty years of process. *Risk Analysis, 15*(2), 137–145.

Fischhoff, B., Lichtenstein, S., Slovic, P., Derby, S. L., & Keeney, R. L. (1981). *Acceptable risk*. Cambridge: Cambridge University Press.

Forrester, I., & Hanekamp, J. C. (2006). Precaution, Science and Jurisprudence: a Test Case. *Journal of Risk Research, 9*(4), 297–311.

Foster, C. E. (2008). Public opinion and the interpretation of the World Trade Organisation's Agreement on Sanitary and Phytosanitary Measures. *Journal of International Economic Law, 11*(2), 427–458.

Freese, W., & Schubert, D. (2004). Safety testing and regulation of genetically engineered foods. *Biotechnology and Genetic Engineering Reviews, 21*, 299–324.

Freudenburg, W. R. (1988). Perceived risk, real risk: social science and the art of probabilistic risk assessment. *Science, 242*, 44–49.

Gabbi, S. (2007). The interaction between risk assessors and risk managers. The case of the European Commission and of the European Food Safety Authority. *European Food and Feed Law Review, 3*, 126–135.

Gaskell, G., Allansdottir, A., Allum, N., Corchero, C., Fischler, C., Hampel, J., et al. (2006). *Eurobarometer 64.3: Europeans and Biotechnology in 2005: Patterns and Trends*, Report to the European Commission's Directorate-General for Research.

Goffman, E. (1974). *Frame analysis: an essay on the organization of experience*. Boston MA: Northeastern University Press.

Goldschmidt, R., Renn, O., Köppel, S. (2008). European citizens' consultations project, Final Evaluation Report. *Stuttgart Contributions to Risk and Sustainability Research*, No. 8. Stuttgart: University of Stuttgart.

Graham, J. D., & Wiener, J. B. (1995). *Risk versus risk*. Cambridge: Harvard University Press.

Gray, P. C. R., Stern, R. M., & Biocca, M. (Eds.). (1998). *Communicating about risks to environment and health in Europe, World Health Organization regional office for Europe in collaboration with the Centre for Environmental and Risk Management, University of East Anglia, UK*. Dordrecht: Kluwer.

Gregory, R., McDaniels, T., & Fields, D. (2001). Decision aiding, not dispute resolution: a new perspective for environmental negotiation. *Journal of Policy Analysis and Management, 20*(3), 415–432.

Grove-White, R., Macnaghten, P., & Wynne, B. (2000). *Wising up: the public and new technology*. Lancaster: CSEC.

Gutteling, J. M., & Wiegman, O. (1996). *Exploring risk communication*. Dordrecht: Kluwer.

Hagendijk, R., & Irwin, A. (2006). Public deliberation and governance: engaging with science and technology in contemporary Europe. *Minerva, 44*, 167–184.

Hammond, J., Keeney, R., & Raiffa, H. (1999). *Smart choices: a practical guide to making better decisions*. Cambridge: Harvard Business School.

Hance, B. J., Chess, C., & Sandman, P. M. (1988). *Improving dialogue with communities: a risk communication manual for government, environmental communication research program*. New Brunswick, NJ: Rutgers University.

Harremoës, P., Gee, D., MacGarvin, M., Stirling, A., Keys, J., Wynne, B., et al. (European Environment Agency) (2001). *Late Lessons from Early Warnings: The Precautionary Principle, 1896–2000*, Environmental Issue Report, No. 22. Luxembourg: Office for Official Publications of the European Communities.

Health and Safety Executive (HSE). (2001). *Reducing risks, protecting people: HSE's decision-making process*. London: HSE.

IRGC (2005). *White paper on risk governance: towards an integrative approach*. Geneva: International Risk Governance Council.

Jaeger, C. C., Renn, O., Rosa, E. A., & Webler, T. (2001). *Risk, uncertainty and rational action*. London: Earthscan.

Jasanoff, S. (1993). Bridging the two cultures of risk analysis. *Risk Analysis, 13*(2), 123–129.

Jasanoff, S. (2005). *Designs on nature*. United States: Princeton University Press.

Jensen, J. K. K., & Sandøe, P. (2002). Food safety and ethics: the interplay between science and values. *Journal of Agricultural and Environmental Ethics, 15*(3), 245–253.

Joerges, C., & Neyer, J. (2003). Politics, risk management, World Trade Organisation governance and the limits of legalisation. *Science and Public Policy, 30*, 219–225.

Joerges, C., & Vos, E. (Eds.). (1999). *EU committees: social regulation, law and politics*. Oxford: Hart.

Johnson, B. B. (1993). Coping with paradoxes of risk communication: observations and symptoms. *Risk Analysis, 13*(3), 241–243.

Joss, S. (1999). Public participation in science and technology policy- and decision-making: ephemeral phenomenon or lasting change? *Science and Public Policy, 26*, 290–373.

Jungermann, H., & Wiedemann, P. (1995). Risk communication–introduction. *European Review of Applied Psychology, 45*(1), 3–5.

Jungermann, H., Schuetz, H., & Thuering, M. (1988). Mental models in risk assessment: informing people about drugs. *Risk Analysis, 8*(1), 147–155.

Kahlor, L. -A., Dunwoody, S., & Griffin, R. J. (2004). Accounting for the complexity of causal explanations in the wake of an environmental risk. *Science Communication, 26*(1), 5–30.

Kasemir, B., Clark, W. C., Gardner, M. T., Jaeger, C. C., Jaeger, J., & Wokaun, A. (2003). *Public Participation in Sustainability Science*. Cambridge, MA: Cambridge University Press.

Kasperson, R. E. (1986). Six propositions for public participation and their relevance for risk communication. *Risk Analysis, 6*(3), 275–281.

Kasperson, R. E., & Palmlund, I. (1988). Evaluating risk communication. In V. T. Covello, D. B. McCallum, M. T. Pavlova (Eds.), *Effective Risk Communication: The Role and Responsibility of Government and Nongovernment Organizations* (pp. 143–158). New York: Plenum.

Kathlene, L., & Martin, J. (1991). Enhancing citizen participation: panel designs, perspectives, and policy formation. *Policy Analysis and Management, 10*(1), 46–63.

Keeney, R., & von Winterfeldt, D. (1986). Improving risk communication. *Risk Analysis, 6*(4), 417–424.
Keynes, J. M. (1921). *A treatise on probability*. London: Macmillan.
Klinke, A., & Renn, O. (2002). A new approach to risk evaluation and management: risk-based, precaution-based and discourse-based management. *Risk Analysis, 22*(6), 1071–1094.
Knight, F. (1921). *Risk, uncertainty and profit*. London: London School of Economics.
Knight, A. J., Worosz, M. R., Todd, E. C. D., Bourquin, L. D. (2008). Listeria in raw milk soft cheese: a case study of risk governance in the United States using the IRGC framework. In O. Renn & K. Walker (Eds.), *Global risk governance* (pp. 179–220). Berlin and Dordrecht: Springer.
Ladeur, K. -H. (2003). The introduction of the precautionary principle into EU law: a pyrrhic victory for environmental and public health law? Decision-making under conditions of complexity in multi-level political systems. *Common Market Law Review, 40*(6), 1455–1479.
Latham, J. R., Wilson, A. K., Steinbrecher, R. A. (2006). The mutational consequences of plant transformation. *Journal of Biomedicine and Biotechnology*, 1–7.
Lehesranta, S. J., Koistinen, K. M., Massat, N., Davies, H. V., Shepherd, L. V. T., & McNicol, J. W., et al. (2007). Effects of agricultural production systems and their components on protein profiles of potato tubers. *Proteomics, 7*, 597–604.
Leiss, W. (Eds.). (1989). *Prospects and Problems in Risk Communication*. Waterloo, ON, Canada: University of Waterloo Press.
Leiss, W. (1996). Three phases in risk communication practice. In H. Kunreuther, P. Slovic (Eds.), *Challenges in risk assessment and risk management: annals of the american academy of political and social science, special issue on risk* (pp. 85–94). Thousand Oaks: Sage.
Leiss, W. (2004). Effective risk communication practice. *Toxicology Letters, 149*, 399–404.
Leiss, W. & Chociolko, C. (1994). *Risk and responsibility*. Montreal, Quebec, Canada: McGill–Queen's University Press.
Lenaerts, K. (1993). Regulating the regulatory process: 'delegation of powers' in the European Community. *European Law Review, 18*, 23–49.
Levidow, L., Carr, S., Wield, D., & von Schomberg, R. (1997). European biotechnology regulation: framing the risk assessment of a herbicide-tolerant crop. *Science, Technology and Human Values, 22*(4), 472–505.
Lipset, S. M., & Schneider, W. (1983). *The confidence gap: business, labor, and government in the public mind*. New York: The Free Press.
Loasby, B. J. (1976). *Choice, complexity and ignorance*. Cambridge, London, New York, Melbourne: Cambridge University Press.
Löfstedt, R. (1997). *Risk Evaluation in the United Kingdom: Legal Requirements, Conceptual Foundations, and Practical Experiences with Special Emphasis on Energy Systems*, Working Paper, No. 92. Stuttgart: Center of Technology Assessment in Baden-Württemberg.
Löfstedt, R. (2001). Risk communication and management in the twenty-first century. *International Public Management Journal, 73*, 335–346.
Löfstedt, R. (2003). Risk communication: pitfalls and promises. *European Review, 11*(3), 417–435.
Löfstedt, R. (2005). *Risk management in post trust societies*. London: Palgrave Macmillan.
Luhmann, N. (1980). *Trust and power*. New York: Wiley.
Luhmann, N. (1989). *Ecological communication*. Cambridge, UK: Polity.
Luhmann, N. (1990). Technology, environment, and social risk: a systems perspective. *Industrial Crisis Quarterly, 4*, 223–231.
Luhmann, N. (1993). *Risk: a sociological theory*. New York: Aldine de Gruyter.
Lundgren, R. E. (1994). *Risk communication: a handbook for communicating environmental, safety, and health risks*. Columbus, OH: Battelle.
Lynn, F. M. (1990). Public participation in risk management decisions: the right to define, the right to know, and the right to act. *Risk-Issues in Health and Safety, 1*, 95–101.
MacIlwain, C. (2005). US launches probe into sales of unapproved transgenic corn. *Nature*. doi:10.1038/nature03570.

Madelin, R. (2007). *How Can we Make Food Safety Governance in Europe More Inclusive?*, Keynote Speech at the Subproject 5 'SAFE FOODS' Presentation Workshop on 11 May 2007. Brussels: Fondation Universitaire.
Majone, G. (2002). Delegation of regulatory powers in a mixed polity. *European Law Journal*, *3*, 330–331.
Marchant, E., & Mosman, K. L. (2004). *Arbitrary and Capricious: the precautionary principle in the European courts*. Washington DC: AEI.
Masson-Matthee, M. D. (2007). *The Codex Alimentarius Commission and its standards*. The Hague: Asser.
Mays, C., Jahnich, M., & Poumadère, M. (2005). A review of institutional arrangements for food safety regulation in France, Deliverable 5.2.5 of Subproject 5 of the EC Framework Programme 6 Integrated Project 'SAFE FOODS'. Cachan: Institut Symlog.
Mays, C., Jahnich, M., & Poumadère, M. (2006). Food safety regulation in France. In E. Vos, F. Wendler (Eds.), *Food safety regulation in Europe: a comparative institutional analysis (Series Ius Commune)* (pp. 217–285). Antwerp: Intersentia.
Miller, H. I., & Conko, G. (2001). Precaution without principle. *Nature Biotechnology*, *19*(4), 302–303.
Millstone, E. (2000). Recent developments in EU food policy: institutional adjustments or fundamental reforms? *Zeitschrift für das gesamte Lebenmittelrecht*, 27(6), 815–829.
Millstone, E., & van Zwanenberg, P. (2002). The evolution of food safety policy-making institutions in the UK, EU and Codex Alimentarius. *Social Policy Administration*, *36*(6), 593–609.
Millstone, E., van Zwanenberg, P., Marris, C., Levidow, L., & Torgersen, H. (2004). *Science in trade disputes related to potential risks: comparative case studies*. Seville: Institute for Prospective Technological Studies.
Moore, C. (1996). *The mediation process. Practical strategies for resolving conflict*. San Francisco: Jossey-Bass.
Morgan, M. G., Fischhoff, B., Bostrom, A., Lave, L., & Atman, C. (1992). Communicating risk to the public. *Environmental Science and Technology*, *26*(11), 2049–2056.
Morgan, M. G., Fishhoff, B., Bostrom, A., & Atmann, C. J. (2001). *Risk communication: a mental model approach*. Cambridge, MA: Cambridge University Press.
Mulligan, J., McCoy, E., & Griffiths, A. (1998). *Principles of communicating risks: the macleod institute for environmental analysis*. Alberta, Canada: University of Calgary.
National Research Council (NRC). (1983). *Risk assessment in the federal government: managing the process*. Washington DC: National Academy Press.
National Research Council (NRC). (1989). *Improving risk communication*. Washington, DC: National Academy Press.
National Research Council (NRC). (1996). *Understanding risk: informing decisions in a democratic society*. Washington DC: National Academy Press.
OECD (2002). *Guidance Document on Risk Communication for Chemical Risk Management*, Series on Risk Management, No. 16, prepared by O. Renn, H. Kastenholz, W. Leiss, Environment, Health and Safety Publications. Paris: OECD.
OECD (2003). *Series on Harmonisation of Regulatory Oversight in Biotechnology, No. 27: Consensus Document on the Biology of Zea Mays Subsp. Mays (Maize)*, ENV/JM/MONO(2003)11. Paris: OECD.
Omenn, G. S. (2003). On the significance of "the Red Book" in the evolution of risk assessment and risk management. *Human and Ecological Risk Assessment*, *9*, 1155–1167.
Owen, H. (2001). *Open space technology*. Stuttgart: Klett-Cotta.
Plough, A., & Krimsky, S. (1987). The emergence of risk communication studies: social and political context. *Science, Technology, and Human Values*, *12*(3–4), 4–10.
Powell, D. A., Blaine, K., Morris, S., & Wilson, J. (2003). Agronomic and consumer considerations for Bt and conventional sweet-corn. *British Food Journal*, *105*(10), 700–713.
Prime Minister's Strategy Unit/UK Cabinet Office (2002). *Risk: improving government's capability to handle risk and uncertainty*. London: UK Cabinet Office.

Public Health Reports special issue on *The Precautionary Principle, 117*(6), November/December 2002.

Ravetz, J. (1999). What is post-normal science. *Futures, 31*(7), 647–653.

Regulation (EC) No 258/97 of the European Parliament and of the Council of 27 January 1997 concerning novel foods and novel food ingredients, *Official Journal of the European Communities*, 14.2.97, No L 43/1.

Regulation (EC) No. 178/2002 of the European Parliament and of the Council of 28 January 2002 laying down the general principles and requirements of food law, establishing the European Food Safety Authority and laying down procedures in matters of food safety, *Official Journal of the European Communities*, 1.2.2002, L 31/1 [General Food Law].

Regulation (EC) No 1829/2003 of the European Parliament and of the Council of 22 September 2003 on genetically modified food and feed, *Official Journal of the European Union*, 18.10.2003, L 268/1.

Regulation (EC) No 1830/2003 of the European Parliament and of the Council of 22 September 2003 concerning the traceability and labelling of genetically modified organisms and the traceability of food and feed products produced from genetically modified organisms and amending Directive 2001/18/EC, *Official Journal of the European Union*, 18.10.2003, L 268/24.

Regulation (EC) No 1935/2004 of the European Parliament and of the Council of 27 October 2004 on materials and articles intended to come into contact with food and repealing Directives 80/590/EEC and 89/109/EEC, *Official Journal of the European Union*, 13.11.2004, L 338/4.

Renn, O. (1999). Diskursive Verfahren der Technikfolgenabschätzung. In T. Petermann & R. Coenen (Eds.), *Technikfolgenabschätzung in Deutschland. Bilanz und Perspektiven* (pp. 115–130). Frankfurt/M.: Campus.

Renn, O. (2004). The challenge of integrating deliberation and expertise: participation and discourse in risk management. In T. L. McDaniels, M. J. Small (Eds.), *Risk analysis and society: an interdisciplinary characterization of the field* (pp. 289–366). Cambridge: Cambridge University Press.

Renn, O. (2007). The risk handling chain. In F. Boulder, D. Slavin, R. Löfstedt (Eds.), *The tolerability of risk. A new framework for risk management* (pp. 21–74). London: Earthscan.

Renn, O. (2008). *Risk governance. Coping with uncertainty in a complex world.* London: Earthscan.

Renn, O., & Levine, D. (1991). Credibility and trust in risk communication. In R. Kasperson, & P. J. Stallen (Eds.), *Communicating Risk to the Public* (pp. 175–218). Dordrecht: Kluwer.

Renn, O. & Walker, K. (2008). Lessons learned: a re-assessment of the IRGC framework on risk governance. In O. Renn & K. Walker (Eds.), *Global risk governance* (pp. 331–367). Berlin and Dordrecht: Springer.

Renn, O., Webler, T., Rakel, H., Dienel, P. C., & Johnson, B. (1993). Public participation in decision making: a three-step-procedure. *Policy Sciences, 26*, 189–214.

Renn, O., Müller-Herold, U., Stirling, A., Dreyer, M., Klinke, A., & Losert, C., et al. (2003). *The application of the precautionary principle in the European Union, Final Report of EU-Project HPV1-CT-2001-00001, PRECAUPRI.* Stuttgart: Center of Technology Assessment in Baden-Württemberg.

Renn, O., Carius, R., Kastenholz, H., Schulze, M. (2005). *EriK – Entwicklung eines mehrstufigen Verfahrens der Risikokommunikation.* In R. F. Hertel & G. Henseler (Eds.). Berlin: Federal Institute for Risk Assessment (BfR).

RISKO, Mitteilungen der Kommission für Risikobewertung des Kantons Basel-Stadt (2000). Seit 10 Jahren beurteilt die RISKO die Tragbarkeit von Risiken. *Bulletin, 3*, 2–3.

Rogers, C. L. (1999). The importance of understanding audiences. In S. M. Friedman, S. Dunwoody, & C. L. Rogers (Eds.), *Communicating uncertainty: media coverage of new and controversial science* (pp. 179–200). Mahwak, NJ: Lawrence Erlbaum.

Rohrmann, B. (1992). The evaluation of risk communication effectiveness. *Acta Psychologica, 81*(2), 169–192.

Rohrmann, B. (1995). Technological risks–perception, evaluation, communication. In R. E. Melchers, & M. G. Stewart (Eds.), *Integrated risk assessment: new directions* (pp. 7–12). Rotterdam: Balkema.

Rohrmann, B., & Renn, O. (2000). Risk perception research–an introduction. In O. Renn, B. Rohrmann (Eds.), *Cross-cultural risk perception: a survey of empirical studies* (pp. 11–54). Dordrecht and Boston: Kluwer.

Rowe, G., & Frewer, L. J. (2000). Public participation methods: a framework for evaluation. *Science, Technology, and Human Values*, *225*(1), 3–29.

Royal Commission on Environmental Pollution (RCEP). (1998). *Twenty-first report: setting environmental standards*. London: Royal Commission on Environmental Pollution.

Sadar, A. J., & Shull, M. D. (2000). *Environmental risk communication: principles and practices for industry*. Boca Raton: Lewis.

Scott, J. (2004). The precautionary principle before the European courts. In R. Macrory (Ed.), *Principles of European environmental law* (pp. 51–74). Groningen: Europa Law.

Scott, J. (2007). *Commentary on the sanitary and phytosanitary measures agreement*. Oxford: Oxford University Press.

Scott, J., & Vos, E. (2002). The juridification of uncertainty: observations of the ambivalence of the precautionary principle within the EU and the WTO. In C. Joerges, & R. Dehousse (Eds.), *Good governance in Europe's integrated market* (pp. 253–286). Oxford: Oxford University Press.

Shackle, G. L. S. (1955). *Uncertainty in economics and other reflections*. Cambridge: Cambridge University Press.

Siegrist, M., Cvetkovich, G., & Roth, C. (2000). Salient value similarity, social trust, and risk/benefit perception. *Risk Analysis*, *20*(3), 353–361.

Sjöberg, L. (2000). Factors in risk perception. *Risk Analysis*, *220*(1), 1–11.

Skogstad, G. (2003). Legitimacy and/or policy effectiveness?: network governance and GMO Regulation in the European Union. *Journal of European Public Policy*, *10*(3), 321–338.

Slovic, P. (1987). Perception of risk. *Science*, *236*(4799), 280–285.

Slovic, P. (1992). Perception of risk reflections on the psychometric paradigm. In S. Krimsky, D. Golding (Eds.), *Social theories of risk* (pp. 117–152). Westport: Praeger.

Spök, A., Hofer, H., Valenta, R., Kienzl-Plochberger, K., Lehner, P., & Gaugitsch, H. (2003). *Toxicological and allergological safety evaluation of GMO*. Vienna: Federal Environment Agency.

STARC (2006). *Current risk communication practices in selected countries and industries*, Deliverable D2, Report by the STARC Consortium to the European Commission. London.

Stilgoe, J., Irwin, A., & Jones, K. (2006). *The challenge is to embrace different forms of expertise, to view them as a resource rather than a burden... the received wisdom. opening up expert advice*. London: Demos.

Stirling, A. (1999). Risk at a turning point? *Journal of Environmental Medicine*, *1*(3), 119–126.

Stirling, A. (2003). Risk, uncertainty and precaution: some instrumental implications from the social sciences. In F. Berkhout, M. Leach, & I. Scoones (Eds.), *Negotiating Change* (pp. 33–76). Cheltenham, UK: Edward Elgar.

Stirling, A., Ely, A., Dreyer, M., Renn, O., Vos, E., Wendler, F. (2006). *A General Framework for the Precautionary and Inclusive Governance of Food Safety. Accounting for Risks, Uncertainties and Ambiguities in the Appraisal and Management of Food Safety Threats*, Working Document produced within Subproject 5 of the EU Integrated Project 'SAFE FOODS', 10 October 2006. Sussex: Sussex University.

Stolwijk, J. A. J., & Canny, P. F. (1991). Determinants of public participation in management of technological risks. In M. Shubik (Ed.), *Risk, organization, and society* (pp. 33–48). Dordrecht and Boston: Kluwer.

Susskind, L. E., Richardson, J. R., & Hildebrand, K. J. (1978). *Resolving environmental disputes. Approaches to intervention, negotiation, and conflict resolution*. Cambridge (Environmental Impact Assessment Project): MIT.

Traavik, T. & Heinemann, J. (2007). Genetic engineering and omitted health research: still no answers to ageing questions. TWN Biotechnology and Biosafety Series 7. Third World Network, Malaysia.

Trettin, L., & Musham, C. (2000). Is trust a realistic goal of environmental risk ommunication? *Environment and Behavior, 32*(3), 410–426.

Turoff, M. (1970). The design of a policy delphi. *Technological Forecasting and Social Change*, 2(2), 84–98.

Trichopoulou, A., Millstone, E., Lang, T., Eames, M., Barling, D., Naska, A., & van Zwanenberg, P. (2000). *European Policy on Food Safety. Final Study*, Working Document for the European Parliament's Scientific and Technological Options Assessment (STOA) Panel, PE 292.026/Fin.St. Luxembourg: European Parliament. www.europarl.eu.int/dg4/stoa/en/publi/default.htm. Accessed 25 May 2004.

Turoff, M. (1970). The design of a policy delphi. *Technological Forecasting and Social Change*, 2(2), 84–98.

UK Advisory Committee on Novel Foods and Processes. (2005). *Paper for Information: ACNFP/72/7 Accidental Cultivation of Bt10 Maize in the USA*, http://www.food.gov.uk/multimedia/pdfs/acnfp_72_7.pdf. Accessed 22 January 2008.

UK Inter-Departmental Liaison Group on Risk Assessment. (1998). *Risk Communication: A Guide to Regulatory Practice*. London. www.hse.gov.uk/aboutus/meetings/ilgra/risk.pdf. Accessed 26 June 2004.

Van den Bossche, P. (2008). *The law and policy of the World Trade Organisation: text, cases and materials*. Cambridge: Cambridge University Press.

Van der Voet, H., de Mul, A., & van Klaveren, J. D. (2007). A probabilistic model for simultaneous exposure to multiple compounds from food and its use for risk–benefit assessment. *Food and Chemical Toxicology, 45*(8), 1496–1506.

Van Gerven, W. (2005). *The European Union. A polity of states and peoples*. Oxford: Hart Publishing.

Van Kleef, E., Frewer, L. J., Chryssochoidis, G. M., Houghton, J. R., Korzen-Bohr, S., Krystallis, et al. (2006). Perceptions of food risk management among key stakeholders: results from a cross-European study. *Appetite, 47*(1), 46–63.

Van Schendelen, M. P. C. M. (Eds.). (1998). *EU Committees as Influential Policymakers*. Aldershot: Ashgate.

Van Zwanenberg, P., & Millstone, E. (2001). Mad cow disease 1980s–2000: How reassurances undermined precaution. In P. Harremoës (Ed.), *Late lessons from early warnings: the precautionary prinicple 1896–2000. Environmental Issue Report*, No. 22 (pp. 157–167). Luxembourg: Office for Official Publications of the European Communities.

van Zwanenberg, P., & Millstone, E. (2005). *BSE: risk, science and governance*. Oxford: Oxford University Press.

van Zwanenberg, P., & Stirling, A. (2003). Risk and precaution in the US and Europe. *Yearbook of European Environmental Law, 3*, 43–56.

Vos, E. (1999). EU Committees: the evolution of unforeseen institutional actors in European product regulation. In C. Joerges & E. Vos (Eds.), *EU Committees* (pp. 19–47). Oxford: Hart.

Vos, E. (2003). Agencies and the European Union. In T. Zwart, L. Verhey (Eds.), *Agencies in European and comparative law* (pp. 113–147). Antwerp: Intersentia.

Vos, E., & Wendler, F. (Eds.). (2006a). *Food safety regulation in Europe: a comparative institutional analysis (Series Ius Commune)*. Antwerp: Intersentia.

Vos, E., & Wendler, F. (2006b). Food safety regulation at the EU level. In E. Vos, F. Wendler (Eds.), *Food safety regulation in Europe: a comparative institutional analysis (Series Ius Commune)* pp. 65–138. Antwerp: Intersentia.

Vos, E., & Wendler, F. (2006c). *A summary report of a workshop with risk managers, for the EC Framework Programme 6 Integrated Project 'SAFE FOODS', contribution to Subproject 5, 2 November 2006*. Maastricht: Maastricht University.

Vos, E., Ni Ghíollárnath, C., & Wendler, F. (2005). *A Review of Institutional Arrangements for European Union Food Safety Regulation*, Deliverable 5.2.6 of Subproject 5 of the EC

Framework Programme 6 Integrated Project 'SAFE FOODS'. Maastricht: Maastricht University.

Webler, T. (1999). The craft and theory of public participation: a dialectical process. *Journal of Risk Research*, *2*(1), 55–71.

Webler, T., Levine, D., Rakel, H., & Renn, O. (1991). The Group Delphi: A novel attempt at reducing uncertainty. *Technological Forecasting and Social Change*, *39*, 253–263.

Webler, T., Rakel, H., Renn, O., & Johnson, B. (1995). Eliciting and classifying concerns: a methodological critique. *Risk Analysis*, *15*(3), 421–436.

Wendler, F., & Vos, E. (2008). *Stakeholder involvement in EU food safety governance: towards a more open and structured approach? Working Paper for Subproject of the EC Framework Programme 6 Integrated Project 'SAFE FOODS'*. Maastricht: University of Maastricht.

Wiedemann, P. & Schütz, H. (2000). Developing dialogue-based risk communication programs, *Arbeiten zur Risiko-Kommunikation*, No. 79. Jülich: Programmgruppe Mensch, Umwelt, Technik am Forschungszentrums Jülich.

World Trade Organisation (WTO) (2004). The Precautionary Principle: Protecting Public Health, the Environment and the Future of our Children. In M. Martuzzi & J. Tickner (Eds.). Copenhagen: WTO.

Wynne, B. (1995). Public understanding of science. In S. Jasanoff et al. (Eds.), *Handbook of science and technology studies* (pp. 361–388). London: Thousand Oaks CA.

Yapp, C., Rogers, B., & Klinke, A. (2005). *A Review of Institutional Arrangements for Food Safety Regulation in the UK*. Deliverable 5.2.2 of Subproject 5 of the EC Framework Programme 6 Integrated Project 'SAFE FOODS'. London: King's College.

Zimmerman, R. (1987). A process framework for risk communication. *Science, Technology, and Human Values*, *12*(3 and 4), 131–137.

Zimmerman, R. & Cantor, R. (2004). State of the art and new directions in risk assessment and risk management: fundamental issues of measurement and management. In T. McDaniels & M. J. Small (Eds.), *Risk analysis and society: an interdisciplinary characterization of the field* (pp. 451–458). Cambridge, MA: Cambridge University Press.

Annex 1
Possible Instruments for Extending Public Participation Beyond the Internet Forum and the Interface Committee

O. Renn

A. Epistemic Discourse

Literature Review and Expert Survey

Reliable risk assessments of simple problems can be undertaken without complicated co-ordination procedures or formal procedures solely on the basis of the available literature or through questions to the corresponding experts. Transparent, plausible presentations of the arguments play a central role when it comes to justifying the results for instance in working groups.

Technical Workshops

Many regulatory agencies and risk assessment institutions frequently hold technical discussions with external scientists or experts. These discussions are aimed at securing the additional information necessary to evaluate the situation and to give external knowledge bearers an opportunity to express their views and arguments. This enables internal experts to gain a comprehensive picture through the exchange of arguments and estimates. The participants get to know the viewpoint of the risk assessment or management agency or other direct players (such as industry or consumer organisations) and source additional information. Technical discussions are not so well suited for resolving conflicts or heated debates. Quite the contrary, under certain circumstances technical discussions may even worsen the tone of a dispute or lead to polarisation.

Expert Hearings

A widespread method of clarifying differences in scientific statements is to invite representatives of the differing views to defend their views to the representatives of the institution (e.g. risk assessment or management agency). The institution representatives put questions to the experts and then give them an opportunity to expand on their arguments. Sometimes open discussions between the experts are also envisaged during the hearings; however, the final decision on how to deal with the dissent lies with the organising institution.

Hearings are excellent and relatively low-cost procedures when it comes to getting to know the diverse opinions of experts and the spectrum of arguments which support every point of view. Hearings do not solve any conflicts nor are they designed to achieve consensus. However, they can create clarity about the underlying reasons which lead to the differing standpoints within a conflict. The authority of the organising institution to take a decision when dealing with dissent depends, firstly, on its sovereign task and, secondly, on the social trust it enjoys. Hearings can certainly improve a situation of trust but they do not suffice in order to give legal validity to decisions.

Expert Committees

Expert committees and scientific committees are also popular tools for involving external knowledge bearers in the safety governance process. They have the advantage over hearings that the experts can communicate freely with one another and that they offer an opportunity for exchanging knowledge and views. They act independently of the public agency or organisation which set them up.

The disadvantages of expert committees are that they do not normally achieve a consensus, require considerable time in order to come to a decision at an unspecified time, are not always able to address the urgent needs of risk managers, and may develop a momentum of their own. Furthermore, expert committees frequently reach agreement only when their members have a similar background and already hold similar points of view. The general public is also extremely sceptical when it comes to the legitimacy of these committees since the criteria for the nomination of the experts are almost always kept secret. Particularly in a high conflict environment, the recommendations of expert committees do not carry very much weight in the eyes of the public at large.

Expert Consensus Conferences and Expert Workshops

In the medical field, experts often come together in a workshop to discuss treatment options and to decide on a generally valid standard (treatment recommendation). The workshop is frequently organised both in working group meetings in order to discuss detailed aspects in depth as well as in plenary meetings in order to obtain general agreement and to elaborate general standards which may be valid worldwide. It might make sense, where statutory provisions permit, to use the tool of an expert consensus conference for the purposes of drawing up and formulating joint agreements for assessments of safety threats.

Delphi Survey

When it comes to priority setting, assessing very uncertain starting situations or highly controversial evaluation results (e.g. in the field of genetic engineering), the classical methods of group work are often overtaxed. In these cases more complex procedures of cognitive judgement are required. One of these procedures, the *Delphi survey*, has proved to be particularly effective. This procedure was developed by RAND Co. in the mid-1960s and initially used for the assessment of defence technologies. Later it was mainly employed as a forecast instrument within the framework of technology impact assessments. The Delphi survey consists of the following steps:

- A research team draws up a catalogue of questions in which the expected consequences of a measure or a decision-making option are examined.
- The questionnaire is sent to a group of recognised experts in the respective field. The experts answer the questions according to the knowledge available to them and estimate the 'subjective certainty', i.e. the estimated validity of their own answers.
- The research team identifies the average values, the extreme values and the variants in the answers.
- The original questionnaire is sent back to the experts together with the evaluation of the first survey. The names of the experts are kept anonymous in order to avoid any influence being exerted by status or seniority. The interviewees are asked to fill out the questionnaire a second time, coupled with the request to use the results of the first survey as a corrective element of their own judgements in their renewed assessment. The purpose of the second survey is to reduce the variance of possible answers and to increase the collective judgement certainty.
- Steps 2, 3 and 4 are being repeated until the experts do not make any further changes to their judgements.

Ideally, the Delphi survey will single out the assessments which are likely to achieve a consensus within the expert group or are the cause of dissent. By anonymising the participants, and through the iterative process of the survey, the respective level of knowledge can be presented without any consideration for the prestige of each of the participants in the Delphi process.

Group Delphi

One of the main disadvantages of the Delphi survey is the lack of substantiation of judgments which deviate from the median of all participants. That is why, together with a few other authors, we have suggested a modification to the procedure, the *group Delphi*. In this case the experts are not linked by means of a postal survey and feedback, but are invited to a workshop lasting between one and two days. What is important here is that the invited experts represent the spectrum of different attitudes and interpretations discussed by the expert world. At the same time, the number of invited experts should not exceed 16–20. In the run-up to or, at the latest, at the beginning of the workshop the task and the structure of the questionnaire should be explained to the participants. Then the participants are divided up into between three and four groups in the first round. Each of these small groups of three to four people is given the same task, i.e. to fill out the questionnaire. The goal is consensus, but deviating votes are possible. In the plenary those experts whose assessments deviate significantly from the mean value of all other participants justify their point of view in front of the others, and defend it in a non-public discourse. The goal of this exchange of arguments is to devote the short time available for communication to those topics for which the greatest discrepancy in estimations has been identified. The goal of the discussions is to establish where the dissent lies, and whether the discrepancies can be overcome through information and arguments from the other experts.

In a second round the procedure is being repeated in new small groups. When putting together the new small groups, care is taken to ensure that representatives of the extreme groups from the first round are spread over all the new groups (permutation of members). The sequence of individual group meetings and plenary meetings is continued until no further significant shifts in standpoints occur. At the end of a group Delphi there is normally a far clearer distribution of answer patterns. The estimates of experts are either scattered around a mean value or they make up multi-peak distributions. In the first case, a consensus has largely been obtained, in the second case there may be several clear separate positions (consensus about the dissent). In both cases the Delphi supplies extensive substantiation for each position.

At the end of this stage one has a profile of suspected or estimated action consequences supported by experts for each decision option for specific criteria. The criteria may also come from the parties involved and, for instance, be elaborated using a prior value-tree analysis. As a consequence of the expert discussions, the verbal substantiations for different assessments are also stored in the profiles as additional information. The disadvantage of the open discussion procedure in the group Delphi is, however, that the participants are no longer anonymous. But prior experience with the group Delphi has shown that status differences have little impact on the group judgement as long as these differences are not dramatic.

The group-Delphi process aims to achieve agreement or non-agreement on cognitive statements. The model is the knowledge discourse based on methodological rules with the goal of identifying apparent dissent, and overcoming this dissent as well as tracing real dissent back to commonly accepted substantiation logics and, by extension, creating consensus via dissent. A discourse of this kind thrives on its exclusivity. If external individuals or representatives of interest groups are actively involved in this discourse, then there is no longer any pressure for methodological substantiation of statements. In most cases, people start strategic positioning in the debate. The discussions frequently end in mutual recriminations, particularly when the experts themselves are polarised in their opinions. At best, observers with no right to vote or speak during the deliberations may be allowed to attend. It is possible to record the discussions with a video camera, too, which makes also sense for the purposes of documenting the course of the discussion. Exclusivity is not a guarantee for the success of a methodologically driven knowledge discourse, but it is at least a necessary precondition. For that reason it is also important to limit the questions to experts to areas of knowledge of relevance for the decision.

Many experts tend to offer political conclusions on the basis of their knowledge as well. One major task of moderation in a group Delphi is, therefore, to prevent an overstepping of the boundaries of collective input knowledge and to remain within the area of the substantiated knowledge of the participants. This is also the only way of keeping to the time schedule of between one and two days.

Surveys and Focus Groups

Surveys of the general public or special groups are excellent settings in which to explore the concerns and worries of the addressed audience. If they are performed professionally, the results are usually valid and reliable. However, the results of surveys provide only a temporary snap shot of public opinion, they do not produce solutions for conflict resolution or predict the fate of positions once they have entered the public arena. Surveys describe the starting position before a conflict may unfold. Focus groups go one step further by exposing arguments to counter-arguments in a small group discussion setting. The moderator introduces a stimulus (e.g. statements about the safety threat) and lets members of the group react to the stimulus and to each other's statements. Focus groups provide more than data about people's positions and concerns; they also measure the strength and social resonance of each argument vis-à-vis counter-arguments. Both instruments provide reliable and valid results for gaining an improved understanding of the context and the expectations of the affected population. They are particularly advisable for input during the stage of concern assessment. But they do not assist the safety managers in resolving a pressing issue (see Annex 1B for citizens' fora and consensus conferences as instruments better suited for this purpose). The major disadvantage of surveys and focus groups is the lack of real interaction among participants. In addition, they are fairly expensive participatory processes.

References for Epistemic Discourse

Benarie, M. (1988). Delphi and Delphilike approaches with special regard to environmental standard setting. *Technological Forecasting and Social Change, 33*, 149–158.

Covello, V. T., & Allen, F. W. (1988). *Seven cardinal rules of risk communication*, OPA-87–020. Washington, D.C.: US Environmental Protection Agency.

Covello, V. T., McCallum, D. B., & Pavlova, M. (1989). Principles and guidelines for effective risk communication. In V. T. Covello, D. B. McCallum, M. Pavlova (Eds.), (1989), *Effective risk communication: the role and responsibility of government and non-government organizations* (pp. 14–24). New York: Plenum.

Dürrenberger, G., Kastenholz, H., & Behringer, J. (1999). Integrated assessment focus groups: bridging the gap between science and policy? *Science and Public Policy, 26*(5), 341–349.

Hance, B. J., Chess, C., & Sandman, P. M. (1988). *Improving dialogue with communities: a risk communication manual for government,* Environmental Communication Research Program. New Brunswick, New Jersey: Rutgers University.
Interdepartmental Liaison Group on Risk Assessment (ILGRA). (1998). *Risk communication. A guide to regulatory practice*. London: Health and Safety Executive.
Krueger, R. A., & Casey, M. A. (2000). *Focus groups: a practical guide for applied research.* Thousand Oaks: Sage.
Lundgren, R. E. (1994). *Risk communication: a handbook for communicating environmental, safety, and health risks*. Columbus, Ohio: Battelle.
Milbrath, L. W. (1981). Citizen surveys as citizen participation. *Journal of Applied Behavioral Science, 17*(4), 478–496.
Mulligan, J., McCoy, E., & Griffiths, A. (1998). *Principles of communicating risks. The macleod institute for environmental analysis*. Calgary, Alberta: University of Calgary, Alberta.
Risikokommission (ad-hoc-Kommission „Neuordnung der Verfahren und Strukturen zur Risikobewertung und Standardsetzung im gesundheitlichen Umweltschutz der Bundesrepublik Deutschland") (2003). *Abschlussbericht*. Munich: Federal Office for Radiation Protection.
Turoff, M. (1970). The design of a policy Delphi. *Technological Forecasting and Social Change, 2*(2), 84–98.
UK Department of Health (1998). *Communicating about risks to health: pointers to good practice*. London: UK Department of Health.
Webler, T., Levine, D., Rakel, H., & Renn, O. (1991). The Group Delphi: a novel attempt at reducing uncertainty. *Technological Forecasting and Social Change, 39*(3), 253–263.
Wiedemann, P., Schütz, H., & Thalmann, A. (2002). *Risikobewertung im wissenschaftlichen Dialog*. Jülich: Forschungszentrum Jülich.

B. Reflective and Practical Discourse

Public Hearings

In many democratic countries, such as the United States, Australia, United Kingdom, France, Switzerland, Germany, and Austria, hearings are statutory components of many approval procedures, regional impact analyses and eco-audits. In the United States, for instance, the Administrative Procedures Act from 1946 stipulates that public hearings must be staged for all projects with major public sector involvement that may have a major impact on the population. Hearings are the most widespread form of structured participation in democratic countries. They are also taking on increasing importance in the Directives of the European Union.

The main advantage of the hearing is the opportunity for a risk assessment or management agency to get to know the worries and concerns of the people affected or the interests of the various groups. In principle, all those concerned are admitted to a public hearing, i.e. the principle of fair representation is upheld. However, practice has shown that it is normally only the activists and representatives of organised interest groups who attend the hearings. In most hearings there are rules for the giving of evidence which only permit factual statements. Finally, hearings are tools for the exchange of information: the stakeholders get to know the views of experts and representatives of public agencies, and the public agency representatives are confronted with the problems and views of the stakeholder representatives.

The rigid rules of the hearing do, however, have some disadvantages. Hearings are normally organized at such a late stage of the risk governance process that they can no longer fulfil their purpose of facilitating a correction should there be serious objections. Because of the limited time and the prerogatives of the panel participants, only a few people have an opportunity to speak. Often lists of speakers are drawn up beforehand or the contributions have to be submitted in advance in writing, which means that spontaneous comments are no longer possible. The equality principle is infringed upon through the division between panel and audience. The participants on the panel normally have special rights (different time limitations). The representatives of the public agencies rarely organize hearings because it is their wish to hear and take on board the concerns of the stakeholders; in general, they merely formally comply with the statutory provisions.

Most empirical studies, therefore, have come to the conclusion that hearings lead to changes in assessments only in very few cases (which does not mean that these changes would always be necessary). Godschalk and Stiftle (1981) examined, for instance, the hearings in North Carolina on water management planning. They came to the conclusion that objections from the groups only influenced decisions in exceptional cases. Irrespective of how open public agency representatives are to objections, the format of the hearing normally leads to a worsening of the conflict rather than to defusing it. The people making the objections know that their only chance to influence the results is by exerting as much public pressure as possible and by flooding public agencies with so many objections that the project can no longer be pushed through politically. The public agency representatives who conduct hearings feel that this reduces them to the role of the fall guys. They scarcely pay any attention to the contents of the objection but do everything they can in order to conclude the procedure in a formally correct manner. This has nothing to do with dialogue. The entire procedure has thus turned into an empty ritual which merely makes the two fronts more entrenched and encourages strategic positioning.

Negotiations Between Important Stakeholders

This form of conflict resolution is predominant in Europe, particularly in the United Kingdom, Germany, and Switzerland, but is also used in the USA under the name of "*Negotiated Rule Making*". The goal of this strategy is to involve the important supra-regional stakeholders in the decision-making process so as to take into account the values and interests of these groups when noting preferences in the decisions. In order to avoid strategic manoeuvring by the participants vis-à-vis the outside world, these negotiations normally take place behind closed doors. Corporatist negotiating strategies of this kind are relatively effective when there is an emergency, and the stakeholders, in principle, agree that action has to be taken.

Where there is no such pressure, then it is normally in the interests of at least one of the participants to keep the process up and running for as long as possible and to delay results until growing public pressure forces a decision. Corporatist solutions, therefore, have three decisive disadvantages:

- Firstly, they exclude all those groups who do not want to, or cannot, comply with the rules of non-public negotiations because they would otherwise lose their clients (example: citizens' action groups).
- Secondly, they only reflect, to a minor degree, the interests and values of the people directly affected by the decisions.
- Thirdly, they lead to a legitimisation deficit in the decision taken because the general public was unable to take part in the decision-making process (lack of transparency). The perception of non-transparency and presumed 'wheeling and dealing' exposes decisions of this kind to public criticism and a lack of acceptance.

The Round Table as a Discursive Procedure

The main goal here is to achieve agreement on the assessment of a given safety threat. Representatives of public agencies and the groups affected by the assessment can have equal rights in the round-table process. A round table begins by specifying the structure of the dialogue and the rights and obligations of all participants. It is the moderator's task to present and explain the implicit rules of the round table to the participants. Furthermore, the participants must jointly lay down decision-making rules, the agenda, the role of the moderator (also with respect to mediation), the sequence of hearings etc. This should always be done according to the consensus principle. All parties must be able to agree to the procedure. There should be unanimous agreement on definitions, possible classifications or other linguistic and comprehension tools. If no agreement can be reached, then the round table must be cut short and postponed to a later date.

Decision-making tools often used in negotiations or round tables include value tree analysis and multi-attribute decision analysis. Those will be described here in brief terms:

Value tree analysis: Once the procedure has been defined, it makes sense to specify the range of statutory foundations (normative statements) which are relevant for the assessment. What is meant here is agreement on the principles which are relevant for the problem in hand. Various methods like the value tree analysis are, in principle, suitable. On the one hand it is necessary to only admit those statements which are closely linked to the topic; on the other hand, for the purpose of fairness, it is necessary to take utmost account of all values and standards which are presented by the respective parties. In this conflict, experience with round tables shows that efforts should be made to record all the values within the framework of conflict mediation, even if the list of values then were to become very long. By contrast, if one reduces discussion to obviously clear values or if one restricts the choices of participants at too early a stage, then some parties will always feel at a disadvantage and re-launch a new 'fundamental debate' at some other stage. In the course of the subsequent negotiations less discriminating values can be excluded.

Multi-attribute utility analysis: Once the values, standards and goals necessary for assessment have been jointly agreed, arguments are exchanged. Four steps can be undertaken to examine the arguments on the basis of analytical decision-making logic:

- Establishment of criteria: A first step involves converting the values and standards accepted by the discourse participants into criteria which directly influence the assessment of the given safety threat (for instance the laying down of the protection good, the determination of the protection goal, the relevant provisions etc.). This conversion must be approved by all participants.
- Validation of knowledge claims: Informed individuals or institutions are asked to assess the evaluation options available according to their best level of knowledge (cognitive correctness). Here it makes more sense to specify a common methodological procedure or a consensus on the experts to be interviewed rather than allowing each group to have its questions answered by their own experts. Frequently, many potential consequences are still contested at the end of this process, particularly when there is a degree of uncertainty. However, the range of possible opinions will be more or less reduced depending on the level of knowledge. Consensus about dissent also helps here to separate controversial from non-controversial claims which promotes further discussion.
- Interpretation: The ranges of expected effects must then be interpreted by the parties. Interpretation means linking factual and value statements to an overall assessment. This assessment can and should be undertaken separately for each aspect of the assessment (for instance, acute health damage, environmental impact etc.). In this way the respective causal chains leading to judgements can be more readily understood. For instance, when interpreting a limit value, the question of trust in the regulatory agency can play an important role. It is then up to the participants to take a closer look at the track record of the respective public agency and, where appropriate, to suggest institutional changes.
- Weighting and weighing up: Even if there were an assessment and interpretation based on common consent, this still would, by no means, mean that there will be agreement. It is far more the case that divergent judgements on decision-making options of the participants can be traced back to different value weightings. In the literature on game theories and economics, this conflict is deemed to be unsolvable, unless one of the participants can convince the others to abandon their preference through the payment of damages (for instance as subsidies), transfer-payments (e.g. a special service) or trade-offs. In reality, however, participants in discussions are indeed open to other participants' arguments (i.e. willing to give up their initial preference) when this loss is still acceptable to them, and, at the same time, the proposed solution is deemed to be 'conducive for the common good', i.e. is considered to be socially desirable in the public perception. If no consensus is reached, then there can and must be a compromise solution which involves negotiating a 'fair' distribution of burdens and benefits.

During a round table the conflicts described here with regard to the procedures, facts, interpretations, and value weightings, must first be identified and then dealt with in a targeted manner through interactive procedures.

Mediation

Mediation procedures involve the bringing in of a neutral mediator for the purposes of conflict resolution and the bringing together of the parties to the conflict, who then will look for solutions in an atmosphere which is conducive to reaching a consensus or, at least, a compromise. In the USA, mediation is closely linked to the model of negotiation and compensation for acceptable disadvantages taken from rational actor theory. The theoretical foundation for mediation is the game theory and its particular application in the negotiation theory as anchored in the so-called Harvard Model. There it is assumed that the entrenched positions of the negotiating partners can be broken down through disclosure of their real interests, and can be turned into a win situation for all those concerned (win–win situation). One good example is the case of two chefs fighting about a lemon. In the course of the dispute it transpires that one of the chefs needs the lemon peel to bake a cake with, whereas the other needs the juice for his tea. So they decide to separate the lemon into juice and peel, rather than splitting it through the middle, and in so doing both parties profit.

Mediation procedures are increasingly gaining a foothold in Europe. The use of mediation is not just about resolving conflicts. Like precautionary health and environmental protection, assessments of safety threats can also be prepared in a participative manner before they escalate into conflicts. The timely bringing together of different attitudes, interests and functions of people at a round table can help.

It is largely up to the moderator to help participants examine the validity of their statements on the basis of previously specified rules. A good moderator has the following characteristics:

- Absolute neutrality in the matter at hand
- Sufficient technical expertise
- Knowledge about statutory rules and provisions
- Expertise and practical experience in chairing discussions
- Social skills in dealing with groups and individuals
- Communication skills
- Focus on the common good, and
- Social respect

Mediation procedures are bound by specific framework conditions. They are suitable for between 25 and 30 people who, in turn, should not represent more than five to ten parties. The participating parties must be able to fall back on a common store of values and goals if there is to be any chance of agreement. Furthermore, it is helpful in the unification process if the parties are already organised and have addressed this topic prior to the procedure.

Although mediation procedures are largely organised on an egalitarian basis and lead to competent judgements, a number of problems remain. The negotiations normally take place behind closed doors which makes it difficult to verify the statements. This has a negative impact on the legitimisation of the results vis-à-vis non-participants. Many analysts are, therefore, of the opinion that mediation is only suitable for those cases in which the knowledge basis has been clearly defined, where the general goals are not disputed and the emotions of the participants play only a minor role. In such cases the different points of view stem from differing interests. For that reason the literature on mediation procedures specifically stresses the use of analytical decision-making or game theory mechanisms for the balancing of interests.

The choice of the rules for discourse management by participants is a major characteristic of the procedure. Even if not all of the parties can participate, it does facilitate at least a representation of the main opponents. The common good can be defended by balancing the possible extremes in the opinions represented. Nevertheless, the lack of participation by unorganised or weakly organised groups continues to be one of the shortcomings of the mediation procedure. That is why they cannot replace the discourse with individuals who are affected but not organised in groups. Furthermore, mediation procedures run the risk of achieving agreement amongst the participating representatives of the invited groups but are often unable to convincingly communicate the solutions to their own members. Hence their members do not feel that they are bound by the negotiated results and may even seek to strip their representatives of power. Without ongoing communication of intermediate results to the members of the groups participating in the mediation procedure, the results of mediation are normally of no further value.

Citizens' Fora (planning cells and citizens' juries)

The involvement of representatives of the public at large in decision-making processes is the main goal of this type of procedure. There is a wealth of different forms which cannot all be looked at individually. Reference is made at this point to all those procedures which diverge from advisory committees in that they give each concerned citizen the same opportunities to participate in the decision-making process. Equal opportunities at local level can be achieved by inviting all those who are potentially affected and facilitating their participation in terms of logistics and time. In the case of more extensive projects, recourse must be made, by contrast, to a selection procedure based on the voluntarism principle or according to a representation method (for instance delegation or random choice). Procedures of this kind aim to ensure that each person concerned has equal chances of participating, irrespective of his/her social position or the degree of organisation of his/her interests.

Two models of citizens' fora have been theoretically elaborated and implemented in practice. Peter Dienel from Wuppertal University has coined the term *planning cell* for these fora. Planning cells are committees of between 10 and 25 people randomly selected who, for a few days, dedicate some of their time to offering decision-making aids on specific questions, and are remunerated for this activity. The underlying philosophy of the planning cell is the desire for fair representation of all those concerned in the preparation and taking of decisions. The planning cell has been used to deal with a number of problems at both local and regional levels.

The second model comes from the Jefferson Centre for Democratic Processes in Minneapolis (U.S. Federal State Minnesota). The founder of the Centre, Ned Crosby, has given his citizens fora the name of '*Citizens'Juries*'. This designation is aimed at highlighting the proximity to juries in the USA. In the same way that jury members use their common sense to determine whether or not an accused person is guilty, the citizens' juries make a recommendation on political options after hearing all the witnesses (experts and representatives of various interests). The model of citizens' juries has been used so far in environmental regulations, educational problems and when electing municipal and regional parliaments in Minnesota.

The legitimacy and efficacy of planning cells or citizens' juries is tied to three preconditions: firstly, the decision-makers must undertake either to accept the recommendations or, at least, to take them into account. Secondly, the organised interests involved in the conflict must agree or, at least, tolerate a mediation solution. This is more likely to happen when the parties no longer perceive any opportunities to resolve the conflict themselves but are more and more convinced that they will be able to present their point of view in a convincing manner to the mediation court. All parties are, therefore, invited to speak as witnesses and present their recommendations. Thirdly, a sufficient number of citizens must be prepared to take on board the obligations linked to participation in the planning cells.

Legitimisation problems are to be expected above all when the population concerned is affected by a measure to very varying degrees. In this case, the people most affected expect to be given more representation in the citizens' fora than they would be allocated by the random principle. Finally, it has been shown that fora, which do not produce any solutions to problems but only indicate approval or rejection of a measure, systematically vote for a refusal because this leads to the fewest internal conflicts within the fora. By contrast, problems which encompass different options with both disadvantages and advantages are particularly suited for citizens' fora. One special advantage of citizens' fora is the opportunity of staging several fora simultaneously in order to address the same issues. This is one way of testing the robustness of the proposed solutions.

The main problems of the citizens' fora are in the area of expertise and follow-up knowledge. Although the fora offer an opportunity to exchange arguments and to use the group dynamics for the assessment of competence, explicit evidence of competence and knowledge are missing. The willingness to listen to experts is no guarantee that factual statements will be examined on the basis of methodological aspects. Nor does confrontation with the preferences of interest groups mean that the appropriateness of the respective values has been examined in any depth. By contrast, citizens' fora offer a good sounding board for anecdotal evidence and statements from day-to-day life, which result from observations or moods. The problems of the competent selection of statements and claims are, therefore, the main thrust of criticism expressed at planning cells, too.

Consensus Conference

The consensus conference model is another innovative method for integrating judgements by lay persons on consumer protection, health and environmental issues into political decision-making processes. The consensus conference consists of the following structural characteristics:

- The discourse organisation, via a newspaper ad, looks for people wishing to participate as lay persons in a consensus conference on a specific subject. Between 10 and 15 people are selected from the interested persons who responded to the ad. In terms of age, gender, education and range of occupations they more or less correspond to a cross-section of the population.
- The selected participants in the consensus conference are given extensive material on the question at stake. The material consists of background reports, newspaper cuttings, expert opinions by the players and other relevant information.
- During two weekends the members of the consensus conference meet for preparatory meetings. At these meetings they exchange their impressions, focus on the main problems, formulate questions for the experts and, with the help of the discourse organisers, select experts to whom they wish to put their questions.

The consensus conference itself is organized on three consecutive days. On the first day the participants put their questions to the invited experts. This is like a classical hearing; the questions are exclusively placed by the participants in the consensus conference. The hearing is public. It is expected that the legal decision-makers (for instance parliamentarians) are present as silent observers. On the morning of the second day the question session can be continued, and questions from the audience may be permitted. In the afternoon the members of the consensus conference come together and prepare a short report with their recommendations. On the third day these recommendations are given to the experts. At a public meeting the experts may provide further information (for instance on factual mistakes or inadmissible generalisations). However, they are not entitled to correct or amend the report. The participants in the consensus conference have another opportunity to finely tune the recommendations in the light of their discussions with the experts. Late in the afternoon of the third day the results are made public and explained at a press conference.

The individual steps in a consensus conference can be further extended or amended. A major component of each consensus conference is the involvement of lay persons as experts in the assessment process and the public hearing with the inclusion of the media and the politically minded public. The procedure has been used mainly in Denmark by the National Board of Technology for problems in regulating genetic engineering, integrated agriculture, risk analyses of chemical additives in foods and also motorised road transport and information technologies. Similar procedures have been used in Norway, Sweden, the United Kingdom, France, Switzerland, Japan and the USA.

Consensus conferences have proved to be a robust, time-restricted and cost-effective variation of discursive decision-making. Prior experience with this tool can mainly be deemed to be positive according to an empirical study by Simon Joss (1997). However, there are a number of problematic points. Participants are chosen using two selection criteria: 'self-selection' by responding to a newspaper and 'outside selection' based on representation criteria by the organisers. Given the low number of selected participants, this is certainly not a representative cross-section of the population. Nor do the advocates of this procedure claim this. But whether the desired heterogeneity in the composition of the participants is sufficient, is questionable despite the best efforts to make a fair selection. Secondly, the influence of individual people cannot be underestimated in a small group. Depending on the composition of the group, the results of the recommendations will be scattered. Hence the legitimisation power of recommendations, particularly in the case of far-reaching collectively binding decision, is difficult to judge. This was also one of the main problems of the first national consensus conference on genetic engineering which was organised by the Hygiene Museum in Dresden.

References for reflective and practical discourse

Agency for Toxic Substances, Disease Registry (ATSDR) (1997). *A primer on health risk communication principles and practices*. ATSDR.

Amy, D. (1987). *The Politics of Environmental Mediation*. Cambridge and New York: Cambridge University Press.

Andersen, S. (1996). Expertenurteil und gesellschaftlicher Konsens: Ethischer Rat und Konsensuskommissionen in Dänemark. In C. F. Gethmann, L. Honnefelder (Eds.), *Jahrbuch für Wissenschaft und Ethik* (pp. 201–208). Berlin and New York: De Gruyter.

Applegate, J. (1998). Beyond the usual suspects: the use of citizens advisory boards in environmental decision-making. *Indiana Law Journal*, *73*, 903–912.

Armour, A. (1995). The citizens' jury model of public participation. In O. Renn, T. Webler, P. Wiedemann (Eds.), *Fairness and competence in citizen participation. evaluating new models for environmental discourse* (pp. 175–188). Dordrecht and Boston: Kluwer.

Breidenbach, S. (1995). *Mediation. Struktur, Chancen und Risiken von Vermittlung im Konflikt*. Köln: O. Schmidt Verlag.

Chekoway, B. (1981). The politics of public hearings. *Journal of Applied Behavioral Science*, *17*(4), 566–582.

Chemical Manufacturers' Association (1988). *Title III community awareness workbook*. Washington, D.C.: Chemical Manufacturers' Association.

Chemical Manufacturers' Association (1994). *Community advisory panel handbook*. Washington, D.C.: Chemical Manufacturers' Association.

Chess, C. (1988). *Encouraging Effective Risk Communication: Suggestions for Agency Management*, submitted to New Jersey Department of Environmental Protection, Division of Science and Research, Trenton, New Jersey, Environmental Communication Research Program. New Brunswick, New Jersey: Rutgers University.

Chess, C., Hance, B. J., & Sandman, P. M. (1988). *Improving Dialogue with Communities: A Short Guide for Government Risk Communication*, submitted to New Jersey Department of Environmental Protection, Division of Science and Research, Trenton, New Jersey, Environmental Communication Research Program. New Brunswick, New Jersey: Rutgers University.

Chess, C., Hance, B. J., & Sandman, P. M. (1989). *Planning Dialogue with Communities: A Risk Communication Workbook*, Environmental Communication Research Program. New Brunswick, New Jersey: Rutgers University.

Claus, F., & Wiedemann, P. M. (Eds.). (1994). *Umweltkonflikte: Vermittlungsverfahren zu ihrer Lösung*. Taunusstein: Blottner Verlag.

Cohen, N., Chess, C., Lynn, F., & Busenberg, G. (1995). *Improving Dialogue: A Case Study of the Community Advisory Panel of Shell Oil Company's Martinez Manufacturing Complex*. New Brunswick, New Jersey: Rutgers University, Center for Environmental Communication.

Crosby, N. (1995). Citizen Juries: one solution for difficult environmental problems. In O. Renn, T. Webler, P. Wiedemann (Eds.), *Fairness and Competence in Citizen Participation. Evaluating New Models for Environmental Discourse* (pp. 157–174). Dordrecht and Boston: Kluwer.

Dienel, P. C. (1978). *Die Planungszelle*. Opladen: Westdeutscher Verlag.

Dürrenberger, G., Kastenholz, H., & Behringer, J. (1999). Integrated assessment focus groups: bridging the gap between science and policy? *Science and Public Policy*, *26*(5), 341–349.

EEI Public Participation Task Force/Creighton, J.L. (1994). *Public Participation Manual* (2nd Edn.). Palo Alto: Edison Electric Institute (EEI).

Fietkau, H.-J. & Weidner, H. (1992). Mediationsverfahren in der Umweltpolitik in der Bundesrepublik Deutschland. *Aus Politik und Zeitgeschichte*, B39-40/92, 24–34.

Fiorino, D. (1990). Citizen participation and environmental risk: a survey of institutional mechanisms. *Science, Technology, and Human Values*, *15*(2), 226–243.

Fisher, R., Ury, W., & Patton, B. M. (1993). *Das Harvard Konzept. Sachgerecht verhandeln, erfolgreich verhandeln*. Frankfurt/Main: Campus.

Folberg, J., & Taylor, A. (1984). *Mediation: a comprehensive guide to resolving conflicts without litigation*. San Francisco: Jossey-Bass.

Gaßner, H., Holznagel, L. M., & Lahl, U. (1992). *Mediation. Verhandlungen als Mittel der Konsensfindung bei Umweltstreitigkeiten*. Bonn: Economica.

Hance, B. J., Chess, C., & Sandman, P. M. (1988). *Improving dialogue with communities: a risk communication manual for government*. New Brunswick, New Jersey: Rutgers University, Environmental Communication Research Program.

Interdepartmental Liaison Group on Risk Assessment (ILGRA). (1998). *Risk communication. A guide to regulatory practice*. London: Health and Safety Executive.

Joss, S. (1997). *Experiences with consensus conferences, Paper at the International Conference on Technology and Democracy, Center for Technology and Culture, University of Oslo, Norway*. London: Science Museum.

Karger, C. R., & Wiedemann, P. M. (1994). Fallstricke und Stolpersteine in Aushandlungsprozessen. In F. Claus, P. M. Wiedemann (Eds.), *Umweltkonflikte: Vermittlungsverfahren zu ihrer Lösung* pp. 195–214. Taunusstein: Blottner Verlag.

Kasperson, R. E. (1986). Six propositions for public participation and their relevance for risk communication. *Risk Analysis*, *6*(3), 275–281.

Leiss, W. (Ed.). (1989). *Prospects and Problems in Risk Communication*. Waterloo, ON, Canada: University of Waterloo Press.

McDaniels, T. (1996). The structured value referendum: eliciting preferences for environmental policy alternatives. *Journal of Policy Analysis and Management*, *15*(2), 227–251.

McKechnie, S., & Davies, S. (1999). Consumers and Risk. In P. Bennett, K. Calman (Eds.), *Risk Communication and Public Health* (pp. 170–182). Oxford: Oxford University Press.

Meyer R. & Sauter A. (1999). TA-Projekt "*Umwelt und Gesundheit" – Endbericht*. TAB-Arbeitsbericht No. 63. Berlin: Büro für Technologiefolgen-Abschätzung beim Deutschen Bundestag.

Morgan, M. G., Fishhoff, B., Bostrom, A., & Atmann, C. J. (2001). *Risk communication. A mental model approach*. Cambridge: Cambridge University Press.

Mulligan, J., McCoy, E., & Griffiths, A. (1998). *Principles of communicating risks*. Alberta: The Macleod Institute for Environmental Analysis, University of Calgary.

National Research Council. (1989). *Improving risk communication*. Washington, D.C.: National Academy Press.

National Research Council. (1996). *Understanding risk: informing decisions in a democratic society*. Washington D.C.: National Academy Press.

Renn, O. (1999). A model for an analytic deliberative process in risk management. *Environmental Science and Technology*, *33*(18), 3049–3055.

Renn, O. (2008). *Risk Governance*. London: Earthscan.

Renn, O. & Oppermann, B. (2001). Mediation und kooperative Verfahren im Bereich Planung und Umweltschutz. In Institut für Städtebau (Ed.), *Kooperative Planung und Mediation im Konfliktfall* (Issue 82, pp. 13–36). Berlin: Deutsche Akademie für Städtebau und Landesplanung.

Renn, O., & Webler, T. (1998). Der kooperative Diskurs–Theoretische Grundlagen, Anforderungen, Möglichkeiten. In O. Renn, H. Kastenholz, P. Schild, U. Wilhelm (Eds.), *Abfallpolitik im kooperativen Diskurs. Bürgerbeteiligung bei der Standortsuche für eine Deponie im Kanton Aargau* (pp. 3–103). Zürich: Hochschulverlag AG.

Renn, O., Webler, T. & Wiedemann, P. (Eds.). (1995). *Fairness and competence in citizen participation. Evaluating new models for environmental discourse.* Dordrecht and Boston: Kluwer.

Sadar, A. J., & Shull, M. D. (2000). *Environmental risk communication. Principles and practices for industry.* Boca Raton: Lewis.

Schneider, E., Oppermann, B., & Renn, O. (2005). Implementing structured participation for regional level waste management planning. In H. S. Lesbirel, S. Daigee (Eds.), *Managing conflict in facility siting. An International Comparison* (pp. 135–154). Cheltenham and Northampton: Edward Elgar.

Striegnitz, M. (1990). Mediation: Lösung von Umweltkonflikten durch Vermittlung. *Zeitschrift für Angewandte Umweltforschung*, *3*(1), 51–62.

Susskind, L. E., & Cruikshank, J. (1987). *Breaking the impasse: consensual approaches to resolving public disputes.* New York: Basic Books.

UK Department of Health. (1998). *Communicating about risks to health: pointers to good practice.* London: UK Department of Health.

Weidner, H. (1995). Innovative Konfliktregelung in der Umweltpolitik durch Mediation: Anregungen aus dem Ausland für die Bundesrepublik Deutschland. In P. Knoepfel (Ed.), *Lösung von Umweltkonflikten durch Verhandlung. Beispiele aus dem In- und Ausland* (pp. 105–125). Basel: Helbing und Lichtenhahn.

Wiedemann, P. M. (1994). Mediation bei umweltrelevanten Vorhaben: Entwicklungen, Aufgaben und Handlungsfelder. In F. Claus, P. M. Wiedemann (Eds.), *Umweltkonflikte: Vermittlungsverfahren zu ihrer Lösung* (pp. 177–194). Taunusstein: Blottner Verlag.

Wiedemann, P. M., Carius, R., Henschel, C., Kastenholz, H., Nothdurft, W., Ruff, F., & Uth, H.-J. (2000). *Risikokommunikation für Unternehmen: Ein Leitfaden.* Verein Deutscher Ingenieure. Düsseldorf: VDI-Verlag.

Zilleßen, H. (1993). Die Modernisierung der Demokratie im Zeichen der Umweltpolitik. In H. Zilleßen, P. C. Dienel, W. Strubelt (Eds.), *Die Modernisierung der Demokratie* (pp. 17–39). Opladen: Westdeutscher Verlag.

Part II
Input and Commentaries by Key Actors in Food Safety Governance

Chapter 11
Input of Key Actors in the Development of the General Framework

M. Dreyer

11.1 Introduction

The General Framework as described in the first part of this book does not result from desk research which took place in academic isolation. It rather reflects the input gained by interviews with and involvement of key actors in the field of food safety governance. One initial source of information were the results obtained through a series of interviews with officials, policy-makers, industry actors, and non-governmental organisations in several EU-Member States and at EU-level. These interviews were conducted for the comparative institutional analysis of food safety regulation in Europe (Vos & Wendler 2006a).[1] From this empirical material important insights were gained into current provisions regarding precaution, participation, the policy-science interface and related reform challenges, and further reforms needed, thus serving as a source of information for the design of the first concept of a General Framework for Food Safety Governance in Europe.[2]

The main methodological pillar in the further elaboration of the governance framework was a *systematic feedback and review process* in form of a series of four workshops, with key actors in the field of food safety governance, at which this first concept was presented and discussed. The workshops were conducted through the autumn of 2006 and involved, successively, industry representatives (Haigerloch/Germany, Castle of Haigerloch, 18–19 September), representatives of non-governmental organisations (London, British Academy, 28–29 September), risk managers (Brussels, Fondation Universitaire, 23–24 October) and risk assessors (Brussels, Fondation

[1] The interview questionnaire was part of a research template that was informed, amongst others, by the outcome of a *consultation process* involving practitioners and scholars in the field of food safety governance: A first draft of the research template was presented to officials from national, European and international food safety institutions and scholars representing diverse and interdisciplinary research areas such as risk and technology, governance and European policy studies at a workshop on "European Food Safety Regulation under Review", held in Stuttgart, Germany, in July 2004.

[2] For this early version of the governance concept, see Stirling et al. (2006).

M. Dreyer and O. Renn (eds.), *Food Safety Governance*,
DOI: 10.1007/978-3-540-69309-3_12, © Springer-Verlag Berlin Heidelberg 2009

Universitaire, 23 November), all of whom were selected to ensure maximum practicable diversity from across Europe.[3] At these workshops important insights were gained into the practicability, and political and social viability of the governance concept.[4]

In particular, the conception of an institutional design of the assessment/management interface as envisioned by the revised General Framework was informed by the outcome of these deliberative events. The review and feedback process was completed on 11 May 2007, when the refined and elaborated governance framework (Dreyer et al. 2007b) was being presented at a final workshop (Brussels, Fondation Universitaire). The objective of this Presentation Workshop was to reflect the amended version with the views of those who had contributed to the feedback process hitherto and with the perspectives, insights and experiences of a wider audience in order to complement the final concept. The present chapter sets out major viewpoints gathered throughout the series of deliberative exercises and it delineates the way in which the earlier version of the governance framework was modified in consideration of this feedback. It goes without saying, that not all of the suggestions and criticisms put forward at the workshops regarding the different elements of the framework architecture and proposed institutional adaptations could be factored into the revision of the governance concept.[5] The mere diversity in views on what was to be considered a critical issue and a possible remedy would have rendered such an undertaking impossible. The revision of the concept was concentrating on those points which were made by several representatives of one actor group and/or also across actor groups. We considered these points to be of particular impact and relevance for the framework's refinement. They are set out in the synopsis at hand. First and foremost the synopsis points out the main lessons that could be learnt from the review and feedback exercise, i.e. that our suggestions for institutional reform had to be reconsidered as far as the following questions were concerned: first, how to achieve a high degree of inclusiveness in the food safety interface activities, and second, how to design structural devices that promise to promote continuity, transparency and accountability in the activities of screening, setting the terms of reference and evaluation without rendering the governance system overly complex and eventually inert.

---

[3]Additional comments were elicited when the framework was presented by Ortwin Renn at a meeting of EFSA's Expert Advisory Group on Risk Communication (Parma, 27 November 2006) and at a meeting of EFSA's Scientific Committee (Parma, 14 December 2006), and by Marion Dreyer at a meeting of EFSA's Stakeholder Consultative Platform (Parma, 26 April 2007).

[4]The discussion technique used at the workshops was based on a sequence of plenary and breakout group sessions. Its purpose was to elicit perspectives specific to the different actor groups as well as to gain insight into main points of consensus and dissent within one actor group.

[5]A more detailed account of the workshop results is provided by the five summary reports produced at the workshops: Dreyer et al. (2006); Ely and Stirling (2006); Vos & Wendler (2006b); Dreyer et al. (2007a); Dreyer and Renn (2007). In each case, these summaries were circulated to the workshop participants to ensure accuracy and to provide the opportunity for further feedback.

11.2 Overall Response

Most of the actor group representatives seemed to agree to the basic assumption underlying the proposed governance framework: The shaping of the interplay between political decision-makers, scientific expert advisors, and corporate and civil society actors throughout the governance process continues to present a major challenge of food safety governance. Ongoing efforts are required for effectively and legitimately coordinating and balancing the involvement undertaken by the different actors. It is in particular the dealing with multifaceted, complex food safety issues and/or cases with high levels of scientific uncertainty where this need is given. The more intense and persistent societal controversies over food production and food safety are usually shaped by these demanding conditions. In this respect, it was noted across the different workshops, that the proposed General Framework would provide interesting ideas and suggestions, some of which were already being developed or implemented – for instance, improved interaction and coordination between risk assessors and risk managers in a system of functional and institutional segregation, or a greater consideration of societal concerns at the different governance stages – yet in a way, which was not very systematic, or at least not as systematic as proposed by the new concept. Critical remarks focused on the institutional reforms, proposed to facilitate the implementation of the envisioned innovative procedures.

11.3 Feedback on Suggestions for Procedural Reform

Most of the actor group representatives generally appreciated the basic architecture of the proposed governance framework as a starting point for further improving food safety governance in Europe. The distinction between the four approaches to assessment and management was considered by most of them as a suitable way of addressing the multiple issues that might be associated with food safety threats, in a more systematic and pro-active manner. In each of the five workshops several participants made the point that these approaches (except for prevention) should not be understood as mutually exclusive but as a set of "tool boxes", each of which would contain devices which may have to be used *in tandem* with those devices of the other tool boxes for dealing appropriately with a given case. This is in agreement also with the governance concept as it was originally designed. The refined account of the concept as presented in this book tries to be more explicit about this provision: Where a given food safety threat displays a number of different challenging attributes, these different aspects may be allocated to parallel treatment by different types of assessment and management (see Sect. 4.2.1). Hence, we do consent that the boundaries between the assessment and management approaches should be considered flexible to a certain degree. The four-approaches concept should not (inadvertently) lead to an inappropriate narrowing of the approaches to assessment (and later on management) with food safety cases cutting across the specified key challenges. We acknowledge that seeing the different assessment and management approaches as potential tools to be used, rather than rigid templates, may help to avoid an inadequate limitation of the scope of the assessment exercise and/or management process.

Most of the actor group representatives agreed about the value of performing a concern assessment in specific cases. In accordance with the intention of the proposed governance framework it was stressed by several of them that the purpose of concern assessment should clearly refrain from representing special interests or offering conflict resolution of value-laden controversies. The revised account of the governance concept makes the objective of this approach to assessment more explicit. It specifies that concern assessment is not about deliberating around values but about gathering social facts and investigating risk perceptions and providing those responsible for evaluation and management with a broader basis of scientific information (cp. Sect. 7.3.2). Also in accordance with the intention of the earlier version of the governance framework many workshop participants underlined that both concern assessment and precautionary assessment[6] were only required under *specific circumstances*: While "routine" cases could be sufficiently dealt with by "standard risk assessment", only specifically challenging cases required these more onerous approaches. This is another feature of the governance concept which we made more explicit when revising the concept's account.

There was also a general appreciation from most actor group representatives of devoting more attention to the interface activities of *framing* and *evaluation*. Several workshop participants agreed that these were essential activities in food safety governance. It was stressed that their establishment as governance steps on their own was a promising way to enhance transparency in the balancing of diverse views and values, which was pointed out as an inherent element of the governance process. It was also acknowledged by many actor group representatives that the interaction between assessors and managers at these stages is particularly important, and that there is room and also preparedness for improving this interaction. It was remarked that, both the European Commission and EFSA have recently increased their efforts in promoting an appropriate and effective working interface and enhanced their cooperation regarding the drafting of the terms of reference of the requests of scientific opinions that the Commission addresses to EFSA. In accordance with the proposed governance concept it was emphasised that it is vital to allow for improved assessment–management interaction *without compromising* the functional differentiation between activities aimed at "understanding" risks and activities aimed at "acting" on risks. It was also considered essential that the relationship and way of coordination between assessors and managers should be open and transparent to all stakeholders as to let them see that this differentiation is being maintained.

---

[6] Very different views were expressed at the workshops on the value of "precautionary assessment". While there was much support by the NGO representatives of this assessment approach, some of the risk managers argued that all precautionary approaches should be left to the risk management stage. Several of the risk assessors and industry experts disputed the distinctiveness of this assessment approach; they considered precaution rather an elaborate and integral part of "conventional" risk assessment.

Several workshop participants agreed that key stakeholders, such as consumer associations and producer organisations, could make a contribution to the conduct of setting the terms of reference and evaluation. While most workshop discussants seemed to affirm the project team's focuses of attention and its diagnosis of the functional need to improve the interaction between actors from politics, science, industry, and civil society, the views diverged on the proposal to institutionalise the interaction between these actor groups at the stages of framing and evaluation through a committee structure (an "Operational Committee", see Sect. 11.4 for more detail).

11.4 Feedback on Suggestions for Structural Reform

While most actor group representatives seemed to agree with our diagnosis of the most important challenges and functional needs in food safety governance, some concern was expressed with regard to the institutional devices which we initially had recommended as possible means to facilitate the implementation of the innovative procedures. Many argued that the proposed introduction of new bodies would add complexity to an already highly convoluted governance system and could end up in bureaucratic overload and undue delays of regulatory processes. Especially, the envisioned introduction of a committee structure for the conduct of the interface activities of setting up the terms of reference and evaluation, met with this type of criticism. The earlier version of the General Framework had envisioned three different options of creating a food safety interface structure. These options differed in the degree of formalisation and included the establishment of an "Operational Committee" (proposed in two slightly different forms) to be composed of assessors, managers, and stakeholder representatives, and as a third option, a more flexible, ad hoc consultation procedure under the auspices of the European Commission (cp. Stirling et al. 2006). While most actor group representatives supported the idea of improving consistency and transparency in the interface activities, and several agreed that a certain formalisation of the framing and evaluation steps could be an appropriate means to this end, many expressed reservations towards the idea of creating a *standing committee* to deal with *all* food safety cases. We had proposed this institutional device as the preferred option of providing the assessment/management interface with an institutional structure. All actor groups expressed their fear that the introduction of this interface structure might result in overall governance structures being too complex, thus entailing undue delays in regulatory processes. In this context, it was stressed by many discussants that they would prefer an interface structure capable of dealing efficiently with the many cases of food safety governance by "bundling up" some cases and leaving out those not requiring in-depth discussion between assessors, managers, and stakeholder representatives.

Moreover, many of the workshop participants disputed the possibility of appointing a limited number of stakeholder representatives for the proposed committee in a manner recognisable as *legitimate*, while keeping the number sufficiently small as not to overstretch the size and operational capacity of the new body. "How to choose the right people" was considered a major issue, and also the question of how to ensure a sufficient representation of the *diversity* of values and perspectives that are usually involved in food safety issues. In addition, several representatives of the consulted groups who considered a standing committee a feasible institutional option stressed that the constitution of the membership of the committee, in particular, and the modalities of stakeholder engagement in the General Framework, in general, would have to be dealt with as issues of democratic legitimacy and power relations.

These critiques prompted us to give greater thought to the institutional adaptations that might facilitate putting the procedural reforms into practice. Thus, we re-considered our recommendation for a preferable institutional design of the assessment/management interface in the light of the two major concerns set out above. The preferred institutional variant of the revised governance concept is the Internet Forum in combination with the Interface Advisory Committee (referred to as the *intermediate proposal,* see Sect. 6.4.2). This variant was designed to improve continuity, transparency, and accountability of the interface activities *and* to, simultaneously, lower the risks of bureaucratic overload and stakeholder involvement restricted to the "Brussels establishment". Through the Internet Forum this institutional option includes a provision for facilitating a higher degree of inclusiveness at all stages of the governance process. We acknowledge that exaggerated and unrealistic claims and aspirations concerning representativeness of the Interface Committee should be avoided. The number of members must be restricted in order to ensure effective working structures. Moreover, judgements over what constitutes the appropriate partitioning of relevant perspectives will depend on the specific context of a given case. The difficulties with the representativeness of the Interface Committee could be alleviated somewhat by combining it with the Internet Forum. We consider this online function – which we propose as the minimum structural reform – a promising means regarding the challenges of feeding a greater diversity of voices (including a wider range of government, scientific expert, academic, commercial industry, and civil society organisations) into the governance process. The Interface Advisory Committee would be requested to deliberate and reflect over the discussions within the Internet Forum as part of its own process of deliberation.

The particular mandate of the Interface Advisory Committee also responds to the concerns about overloading the governance process. It works in an advisory function only and deals merely with selected cases. Also the possibility of "bundling up" cases is meant to enhance the effectiveness of the working of this body. The institutional device of a standing committee with responsibility for all cases (this had initially been the preferred institutional option and is now denoted the "Interface Steering Committee") in combination with the Internet Forum is referred to as the "maximum proposal" in the revised concept. This terminology is meant to account for the fact that this is the institutional variant with the broadest mandate which had met with some criticism in the feedback process.

Furthermore we re-considered our initial proposal for a structural device for the screening step. Our revised recommendation for a structure to assist the fulfilment of the screening function, i.e. the tailoring of the assessment exercise to key attributes of food safety threats (cp. Sect. 4.2), also reflects the concerns expressed over institutional changes that could result in too complex a governance structure. The earlier version of the General Framework envisioned the creation of a Screening Board with full responsibility for this governance activity. From several workshop participants' point of view, this Board would add a major, yet unnecessary bureaucratic layer to the governance system. Screening activities, it was noted, have already been performed in the current governance system by EFSA's scientific panels, albeit in an informal and ad hoc manner. The Screening Unit which the revised General Framework envisages would not conduct the investigation of the screening questions itself. Its mandate is rather to act as a *clearing house* between the secretariat of EFSA and the various scientific panels at the stage of screening. It would co-ordinate the referral of screening questions to the Scientific Panels and expert services, and the collection of the answers from the respective scientific units (see Sect. 6.3).

11.5 The Revised Recommendation for Designing the Food Safety Interface Under Review

As set out above, it was mainly our suggestions for structural reform – and here the proposed Interface Committee in particular – which met with reservations and criticism. As the next chapter will show, our revised recommendation for organising the food safety interface does not (fully) alleviate the two main concerns raised in the feedback process either, i.e. concerns about bureaucratic overload and insufficient inclusiveness and representativeness of the Interface Committee. The point of the present subsection is to sketch and highlight the different views the four commentators, who had been invited by us to contribute a written review of the revised version of the General Framework, take on our modified recommendation.

All four commentaries welcome the suggestion for further enhancing the use of the Internet as a means of engaging with a wider range of social groups, in the official processes of handling food safety threats. The Internet Forum – the food safety interface institution which the General Framework recommends unconditionally – is considered a promising idea in this respect, provided that, as pointed out by the consumers' association representative, the provisions and efforts to be made will ensure that this online function does not end up in a forum for debate among the group of "usual suspects" (Davies, this volume, Sect. 12.3, p. 227), i.e. those disposing of larger resources and powerful positions and having been part of the "Brussels establishment" for a long time. This relates to the difficult issue of "How to engage the unengaged".[7] The Internet Forum, we claim, is a possible means for having a greater diversity of views and values represented in the governance process.

[7] With reference to the recommendations emerged from the DG SANCO 2006 Healthy Democracy Process (European Commission, February 2007), Director General Robert Madelin called for effective solutions to "engage the unengaged" in his keynote speech at our Presentation Workshop in Brussels on 11 May 2007 (Madelin 2007); Atkins and Norman also point out this challenge in their commentary in this volume, Sect. 12.1, p. 211.

However, we admit that it will require more than the technical facilities to achieve this aim. There are substantial challenges involved in terms of enabling and encouraging all those interested and affected to participate in the envisioned online deliberations and consultations. These challenges and possible ways of meeting them deserve further reflection and discussion.

All of the commentators also support the recommendation to further develop the interaction between assessors and managers at the stages of framing and evaluation. All commentaries, however, continue expressing their concerns – to a varying extent – as to formalising this interaction through a committee structure. We understand two commentaries as regarding the advocated Interface Advisory Committee as a principally positive proposal, while stressing specific prerequisites for reaching the aims that are attached to this new structure. One of these commentaries, in relation to both proposed interface institutions, cautions that they ought to be subjected to more general provisions for avoiding an "overkill of participatory procedures" (Noteborn, this volume, Sect. 12.2, p. 220). The other commentary warns that reflection and discussion would be needed on how to make certain that both interface institutions, and the Interface Committee in particular, will not "inadvertently make the risk analysis process more closed and exclusive" (Davies, Sect. 12.3, p. 231). The other two commentaries take a more critical stance on the proposed committee structure. With regard to the assessment–management interface they express a clear preference for further developing existing relationships and fora. One of these commentaries cautions that even the proposed Interface Advisory Committee seeing only particularly challenging cases "may prove onerous unless there are very clear guidelines as to what constitutes a challenging case" (Atkins & Norman, Sect. 12.1, p. 212). The other commentary expresses even stronger reservations about the advocated interface committee pointing to the issue of representativeness and alerting to "risks to the efficiency and timeliness" of the risk analysis process (Rawling, Sect. 12.4, p. 238).

It becomes (even more) apparent from this second-stage feedback that the limited number of stakeholders sitting on the Interface Advisory Committee (for the sake of practicable working structures) invites legitimate questions about inclusiveness and representativeness. We acknowledge that the proposed new structure will face particular justification requirements in this respect. In our view, substantial challenges deserving more reflection and research lie in, first, effectively linking the two advocated food safety interface structures in a way which ensures that the Interface Committee accounts for the input and output of the Internet Forum which was designed to add to the inclusiveness of all four major stages of the governance process. We concede that a second major challenge will be to enable and encourage a wide range of social actors from all over the European Union to contribute to the Internet-based consultations and deliberations. The Interface Advisory Committee, we believe, could actually be helpful in this respect. Vesting framing and evaluation activities with a "face" could lead to a perception of agency in relation to these two stages at which the integration of scientific and socio-political and socio-economic considerations are of utmost importance and, therefore, motivate engagement via the Internet Forum. We recognize that clear guidelines on what constitutes *particularly challenging cases* would be the key to implementing the Interface Advisory Committee in a way which does not produce undue delays in the whole governance process. The setting up of these guidelines would be an essential task to be dealt with at the review stage (cp. Sect. 3.2). The concepts of uncertainty and ambiguity and the proposed screening criteria could serve as a basis for this undertaking.

References

Dreyer, M. & Renn, O. (2007). *A Summary Report of the Presentation Workshop*, for the EC Framework Programme 6 Integrated Project 'SAFE FOODS', contribution to Subproject 5, 23 July 2007. Stuttgart: DIALOGIK.

Dreyer, M., Renn, O., & Borkhart, K. (2006). *A Summary Report of a Workshop with Industry Representatives*, for the EC Framework Programme 6 Integrated Project 'SAFE FOODS', contribution to Subproject 5, 19 October 2006. Stuttgart: DIALOGIK.

Dreyer, M., Renn, O., & Borkhart, K. (2007a). *A Summary Report of a Workshop with Risk Assessors (EU-Member State level)*, for the EC Framework Programme 6 Integrated Project 'SAFE FOODS', contribution to Subproject 5, 15 January 2007. Stuttgart: DIALOGIK.

Dreyer, M., Renn, O., Ely, A., Stirling, A., Vos, E., & Wendler, F. (2007b). *A General Framework for the Precautionary and Inclusive Governance of Food Safety*, for the EC Framework Programme 6 Integrated Project 'SAFE FOODS', Interim Report of Subproject 5, 4 May 2007. Stuttgart: DIALOGIK.

Ely, A. & Stirling, A. (2006). *A Summary Report of a Workshop with NGO Representatives*, for the EC Framework Programme 6 Integrated Project 'SAFE FOODS', contribution to Subproject 5, 2 November 2006. Brighton: University of Sussex.

Madelin, R. (2007). *How Can We Make Food Safety Governance in Europe More Inclusive?*, Keynote Speech at the Subproject 5 SAFE FOODS Presentation Workshop on 11 May 2007, Brussels: Fondation Universitaire.

Stirling, A., Ely, A., Dreyer, M., Renn, O., Vos, E., & Wendler, F. (2006). *A General Framework for the Precautionary and Inclusive Governance of Food Safety. Accounting for Risks, Uncertainties and Ambiguities in the Appraisal and Management of Food Safety Threats*, Working Document produced within Subproject 5 of the EU Integrated Project 'SAFE FOODS', 10 October 2006. Sussex: Sussex University.

Vos, E. & Wendler, F. (Eds.) (2006a). *Food Safety Regulation in Europe: A Comparative Institutional Analysis* (Series *Ius Commune*). Antwerp: Intersentia.

Vos, E. & Wendler, F. (2006b). *A Summary Report of a Workshop with Risk Managers*, for the EC Framework Programme 6 Integrated Project 'SAFE FOODS', contribution to Subproject 5, 2 November 2006. Maastricht: Maastricht University.

Chapter 12
Commentaries on the Revised General Framework[1]

12.1 A Risk Management Perspective on the Governance Framework

Commentary from Dr. David Atkins and Dr. Julie Norman

12.1.1 Introduction

1. All of us occupy private realms of unreason where we merge evidence, myth, and belief. It is part of the human condition that determines who we are as individuals; but in public life the transparent use of evidence and robust scientific analysis are paramount to winning trust in decisions taken by experts and politicians. Nowhere is this more important than in the fields of health and food safety, where those involved have a duty to be clear about how judgements have been reached and to be explicit about how the different streams of evidence and analysis have contributed to the outcome.
2. Fundamental to this duty is good science governance, based on agreed and transparent best practice, that sets out for both the expert and lay stakeholder how evidence and scientific analysis are used in the decision-making process. Operated transparently such governance makes the decision-making process accountable, open to challenge and review, and is fundamental to winning the trust of stakeholders in the advice and decisions of experts. This principle is at the heart of the welcome proposal for 'a General Framework for the Precautionary and Inclusive Governance of Food Safety'.

[1] Footnote by the editors: It is important to note that the report version that was subjected to commenting (Dreyer et al. 2007a) did neither include the chapter on risk communication, nor the chapter presenting the case study on genetically modified maize. These two chapters have only been added to the present volume (Chaps. 8 and 9). The four invited commentaries, therefore, do not relate to them. Neither were the chapters part of the documentation that was used during the series of workshops with key actors in food safety governance. Hence, they did not form part of the feedback and review process.

M. Dreyer and O. Renn (eds.), *Food Safety Governance*,
DOI: 10.1007/978-3-540-69309-3_13,

12.1.2 Interface Relationships in Risk Analysis

3. The classic risk analysis model has been set out clearly by Codex (see World Health Organisation (WHO) and Food and Agriculture Organisation of the United Nations (FAO) 2006). It is, however, helpful to set out the stages in a bit more detail to emphasize that the integrity of science governance relies on applying best practice throughout the risk analysis journey:

- Framing the question for risk assessment
- Collecting the evidence
- Analysing the evidence
- Considering the nature and extent of uncertainty
- Formulating the risk assessment advice
- Explaining the risk assessment conclusions
- Developing risk management options
- Deciding on the best policy and advice
- Communicating it effectively
- Measuring its impact and reviewing its effectiveness.

These steps will seem familiar and we imagine that all regulatory authorities ensure that they are carried out to some extent. However, it is often hard to distinguish the different steps, see how they inter-relate and how each contributes to the quality of the overall outcomes.

4. It is now accepted that the integrity of the risk analysis outcome depends upon functional separation of responsibility for risk assessment and risk management. This principle protects the scientific integrity and independence of the scientific risk assessment by ensuring that it is not influenced by the policy preferences of risk managers. This need for functional separation is the basis for establishing the European Food Safety Authority (EFSA). However, it is also true – and this clearly emerged from the workshop discussions – that efficient risk analysis also requires effective communications between risk assessors and risk managers which are transparent and accessible to stakeholders.
5. The UK Food Standards Agency differs from EFSA and other agencies because we carry out both risk assessment and risk management activities. However, we do still maintain a functional separation between risk assessment – which is carried out by independent scientific advisory committees (SACs) which are equivalent to EFSA Panels – and risk management – which is carried out by the Food Standards Agency's Board. This has given us better opportunities to establish a dialogue between the two functions and has also meant that we have had to give careful thought to governance issues to maintain our credibility as an open and trustworthy body. We have learned that risk assessors need to understand how risk managers work so that they can address the right questions and present their opinions in ways that the risk managers will understand and find useful. It is essential that this relationship is open and transparent to all of our

stakeholders so that they see that the integrity and independence of the risk assessment are not compromised.

6. The Agency has begun to investigate how stakeholders can be involved in framing the questions for risk assessment, and we are very pleased to see the specific reference in the General Framework to seeking a social science input. We agree that social science has an increasing role to play in supporting risk assessment and risk management. The Agency and the Royal Society (the UK's academy for science) held a workshop in 2005 (The Royal Society & Food Standards Agency 2005) to consider the influence of social and institutional assumptions in assessing, managing and communicating risk particularly in cases where a high degree of uncertainty exists. We hope that further dialogue about how to use social science will take place under the auspices of EFSA. Another challenge – as the General Framework rightly points out – is dealing with uncertainty. The FSA expects the risk assessment to define the uncertainties clearly for the risk managers to enable them to respond effectively, giving clear messages about what is known, what is unknown and what they are doing about it.
7. When undertaking the risk management part of the business, a far wider evidence base needs to be drawn upon, e.g. individual liberty, regulatory constraints and the feasibility of actions to manage risks, economic and social consequences, and consumers' appetite for risk. This second stage carried out by the Commission and Member States, and in the UK by the independent Board of the FSA, is distinct from the scientific process of advocacy and challenge that generates the risk assessment. Ideally it is an iterative, consultative process leading to accountable, transparently achieved judgments. Scientific risk assessment cannot 'prove' safety, but is the starting point for judgements on risk management. Reviewing the effectiveness of decisions and their implementation benefits the FSA's performance as managers of risk. It is essential to be consistent and proportionate, and base recommendations and actions on the balance of risks and benefits to everyone concerned.
8. The FSA has found that the application of open and transparent science governance processes has helped to develop effective engagement with stakeholders throughout the risk analysis journey. Although there is scope for improvement, we have found that clarity about the science governance processes has proved fundamental to enabling stakeholders to see and challenge that best practice has been followed and to engage with the process. You will see from Fig. 12.1[2], describing this journey, that the FSA actively engages with stakeholders at all key stages. This provides a valuable external challenge that gives greater assurance when making policy decisions. It also provides vital feedback on the effectiveness of advice and policy that helps us to improve.

---

[2] Cp. www.food.gov.uk/multimedia.pdfs/fsa060207.pdf. Accessed 31 January 2008.

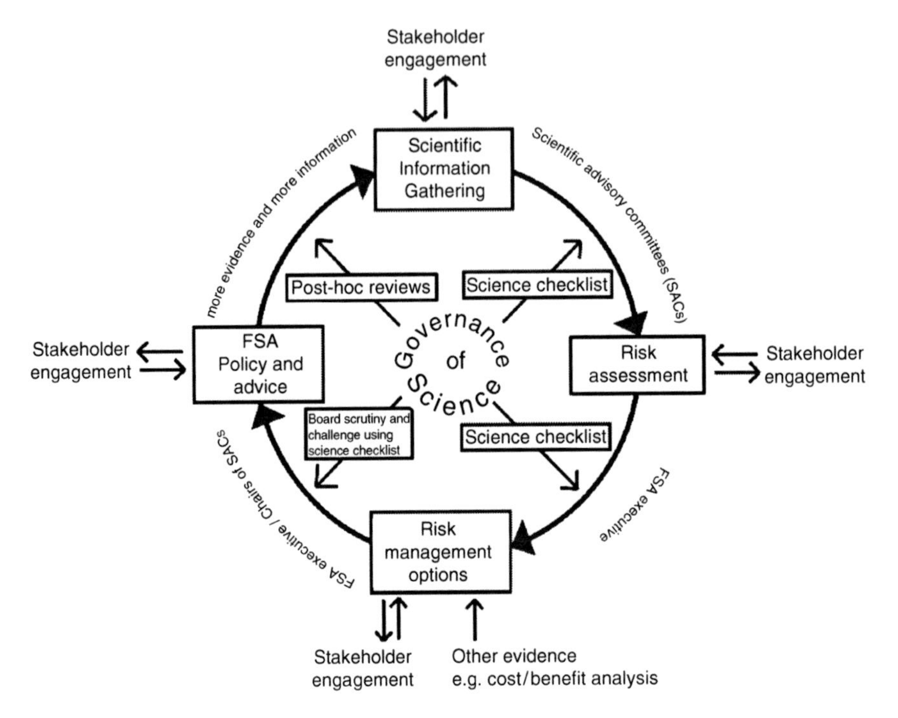

Fig. 12.1 FSA engagement with stakeholders

9. In the Agency's view, one of the strengths of this model is that both the risk assessment and risk management parts of the process are equally open. We see as undesirable a situation where the risk assessment opinion is published and available for challenge, but the risk management discussion is held in private and it is not clear what other factors are being taken into account.
10. While recognising that the independence and integrity of the scientific risk assessment must not be compromised by risk management considerations, we do not believe that risk assessment, management and communication can or should be totally separate, sequential activities. Our open and transparent approach has moved us away from the sequential model of 'assess, decide, tell' towards a more integrated model that has engagement with the public and with other stakeholders at its core.
11. One step which we have found particularly useful is to bring together formally the risk assessors and risk managers as part of advice/policy development. The FSA has 10 expert scientific advisory committees (SACs; the equivalent of the EFSA Panels). We use these committees selectively: straightforward risk assessment is often carried out by our in-house scientists, but we seek a committee opinion where it is essential to have an independent view. This might be because there is uncertainty or because key risk management decisions will be based on it. Until two years ago, there was little contact between these

SACs and the FSA Board which makes the risk management decisions. This has now changed with the Chairs of the SACs being invited to answer the Board's questions at open meetings. We are careful to ensure that the SACs are not becoming involved in the risk management decisions. However, it is useful for the SACs to gain direct experience of the sort of questions that risk managers ask and the language that they speak. This has greatly facilitated the interchange and understanding of views.

12.1.3 Critique of the General Framework

12. The proposal for 'a General Framework for the Precautionary and Inclusive Governance of Food Safety' is a welcome contribution to the debate with the importance it places on transparent engagement at the key interfaces of framing the question and evaluating the risk assessment. While effective communication between EFSA – as the risk assessors – and DG SANCO officials – as the primary risk managers – can be handled through the development of existing relationships and fora, it may be necessary to consider new approaches to engage with other key risk management players including the European Council, the European Parliament and Member States. There is further challenge presented by the diversity of stakeholder groups within the EU, their centralization[3] and their appetite and ability to engage. Engaging with the unengaged will remain a substantial challenge requiring innovative flexible approaches.
13. We think that the discussion which is going on to develop the Framework is essential to develop a shared understanding across Member States and institutions about what constitutes a good governance model. The approach that the authors have taken in clearly separating out four processes (*Framing, assessment, evaluation,* and *management*) rather than the traditional two of risk assessment and risk management provides real clarity and enables people with different interest to see where they can make a contribution. We do share the concerns expressed by some at the workshops that the result of this will be to add time delays and unnecessary bureaucracy to the overall process. However, we do not think this is inevitable.
14. The first step to be supported is the building of capability at EFSA, whether this is through formal establishment of a Screening Unit and Panel on Concern Assessment or by developing that expertise more widely in the organization. The Internet Forum is an excellent idea. It will make best use of technology to involve a wide spread of stakeholders at the different stages, and will potentially benefit consumer groups by enabling their scarce resources at the points in the process which interest them most. It also has the benefit of speed – EFSA will be able to get opinions rapidly if it needs to.

[3] Large federations cannot be expected to reflect the interests of all food sectors in each EU country.

15. The role assigned to the Interface Institutions (of setting the terms of reference for the risk assessment and evaluation) is important. To suggest that all cases are referred to an Interface Steering Committee would introduce quite unacceptable delays and bureaucracy into the process. Even an Interface Advisory Committee seeing only 'particularly challenging' cases may prove onerous unless there are very clear guidelines about what constitutes a challenging case. It is here we see the main value of an Interface Committee – as a body which meets initially to map out the sorts of questions which need to be addressed when setting the terms of reference and producing a guide to the sorts of evidence which will feed into the evaluation. The Agency has developed a 'Science Checklist' (see Annex in Food Standard Agency 2006) which actually performs this general function (though we do not differentiate so explicitly the different stages of the process). Not all factors will apply to every case but having the full range of issues set out acts as an aide-memoire of what needs to be considered. If this initial stage is carried out in a comprehensive manner, it may well be that few cases would not be covered and hence need to be referred to the Interface Advisory Committee. We would recommend that the committee in any case should meet occasionally to learn lessons from how the guidelines are working out in practice.

12.1.4 Conclusions

16. The Framework provides an opportunity to understand exactly what components are needed for a systematic and transparent approach to risk analysis. It brings to light two key processes which have hitherto been hidden at best, or rather neglected at worst: framing the question and considering the implications of the risk assessment outcome before it is passed on to the risk managers (evaluation). Once these processes have been assimilated into the EU food safety system, it will be more robust and stakeholders will have a much clearer understanding and hence increased confidence in the outputs.
17. The main challenge is to incorporate the proposed changes without increasing the time taken to make risk management decisions and without increasing the complexity of the process. We believe that this can be done and will work with EFSA, the Commission and Member States to achieve this aim.

References

Dreyer, M., Renn, O., Ely, A., Stirling, A., Vos, E., & Wendler, F. (2007). *A General Framework for the Precautionary and Inclusive Governance of Food Safety, for the EC Framework Programme 6 Integrated Project 'SAFE FOODS', Interim Report of Subproject 5, 4 May 2007*. Stuttgart: DIALOGIK.

Food Standards Agency (2006). *Annual Report of the Chief Scientist 2006/07*. www.food.gov.uk/multimedia/pdfs/board/fsa071005a.pdf. Accessed 30 January 2008.

The Royal Society & Food Standards Agency (2005). *Social Science Insights for Risk Assessment: Findings of a Workshop held by the Royal Society and the Food Standards Agency on 30 September 2005*. www.royalsoc.ac.uk/downloaddoc.asp?id = 2797. Accessed 30 January 2008.
World Health Organisation (WHO) & Food and Agriculture Organisation of the United Nations (FAO) (2006). *Food Safety Risk Analysis. A Guide for National Food Safety Authorities* (FAO Food and Nutrition Paper 87). www.who.int/foodsafety/publications/micro/riskanalysis06.pdf. Accessed 30 January 2008.

12.2 A Risk Assessment Perspective on the Governance Framework, with a Focus on the Proposed Procedural Reforms of Risk Assessment

Commentary from Dr. Hubert P.J.M. Noteborn

The European Union (EU) has been at the forefront of the development of the risk analysis principles and their subsequent international acceptance. In this vein, the White Paper on European Governance calls for better involvement of the public in order to make the EU system more open and accountable to its citizens (Commission of European Communities 2001). This includes more formal stakeholder participation (engagement), improved transparency of the process and publicly accessible reports at each stage of the open, cyclical, iterative and interlinked process. Therefore, I very much appreciated the 'General Framework for Precautionary and Inclusive Governance of Food Safety' of the SAFE FOODS project. In particular, the proposed four approaches to the assessment of threats using screening criteria for conditions of certain and unambiguous risks (i.e. *preventive measures*), quantifiable risks (i.e. *risk-based assessment*), scientific uncertainty (i.e. *precautionary assessment*) and socio-political ambiguity (i.e. *concern assessment*): a transparent and effective strategy composed of an assessment of adverse effects and other comparative analyses. It includes open dialogues to support the awareness of the risk managers of uncertainties, pros and cons of options, and different other influencing factors. Indeed, notwithstanding the fact that the adoption of Regulation (EC) 178/2002 and the new EU institutional arrangements have proved to be effective and productive in recent evaluations (i.e. performance audit/review of EFSA, August 2005), several aspects of the implemented system deserve more attention and possible improvements in the near future.

Generally speaking, I am satisfied with the functional separation between risk assessment and risk management as established at EU level by Regulation (EC) 178/2002. However, the relationship between political judgment and science-based expertise is a troubled one: living-apart-together. According to Codex Alimentarius (2007), the risk assessment policy should be defined by risk managers in close consultation with risk assessors and other stakeholders. The question of how to understand and design the boundaries between the components of risk assessment and risk management in functional and institutional terms is a subject for current debates. For example, a definition and the principles regarding the Community risk assessment policy is lacking in Regulation (EC) 178/2002, Article 6. Among others, it seems appropriate to improve the interactions between risk assessors and risk managers, and between EU and national levels.

According to Hoppe (2005), three cliché images compete in the media. The business-as-usual political myth is that, in spite of appearances to the contrary, politics is safely 'on top' and experts are still 'on tap'. Whereas scientific experts claim that powerless but inventive scholars only 'speak truth to power', cynics even state that the scientific advisers would follow their own interests, unless better paid by the interests of politicians or industrialists; because these stakeholders would ask the 'hired guns' for advice only to support and legitimize, respectively, their preformed political decisions or established business interests. To the extent this cynical perspective gains ascendancy, safety politics and risk science lose credibility (Hoppe 2005). Although none of these cliché images would stand firm on closer investigation (Hoppe 2005, 2007), developments in EU regulation of food, sanitary and nutritional safety are directed to a functional separation of risk assessment and risk management (Commission's DG SANCO 2005; EFSA 2006). There is a desire to separate the scientific consideration of risk from the broader task of weighing risk against benefits and other factors. These include, for example, the feasibility of controlling a risk, the most effective cost–benefit actions depending on the part of the food supply chain where the problem and the socio-economic effects occur. The primary motivation expressed is to ensure that the risk assessment is independent and not influenced by outside interests such as industrial sponsors or pressure groups, nor by policy considerations. However, nearly every publication favouring separation emphasizes also the need for an efficient interaction. This ensures the assessment output to be fit for purpose, and to optimize the use of resources and time. Like Funtowicz, Sheperd, Wilkinson, and Ravetz (2000), I advocate the emergence of 'post-normal science' and have added to this an extended peer-review. Moreover, the envisaged improvements should guide expert advisers and policymakers as well as other stakeholders in their day-to-day boundary work.

The challenge is to develop a pragmatic approach of sophisticated images of boundary arrangements (i.e. interfaces) between risk assessment and risk management. It should achieve both functional separation and efficient interaction as enshrined in the General Food Law (Regulation (EC) 178/2002). It can be argued that the outcomes of the SAFE FOODS project prepare successfully the ground here. In addition to a more comprehensive risk assessment, as appropriate to the circumstances (Regulation (EC) 178/2002, Article 7), the project team suggests a procedural and institutional reform by introducing interfaces into the risk analysis paradigm of Codex Alimentarius: mandating (or framing) and evaluation; and discursive processes of an active deliberation of all key players (e.g. risk assessors, risk managers, communicators, industrialists, scientists, and consumers): on the one hand to define the subjects down for consideration (problem definition and context) and, on the other, to process the outcomes of risk assessment for evaluation (conclusion and appraisal) and decision-making (selection) on appropriate instruments to mitigate risks. Indeed, their research and recommendations would help to address what society wishes them to take care of (e.g. Regulation (EC) 178/2002, Article 22(4) and 23(f)). Participatory interacting at the very start of defining the objectives (i.e. framing and screening) allows drawing a planning process and strategy to any sensitive or controversial aspect at stake (Regulation (EC) 178/2002, Article 9; Consultation of EFSA according to Article 29 or specific other EU Regulations). These elements shape the question(s) for assessment (or Terms of Reference as defined in Regulation (EC) 178/2002, Article 3).

Key is also a transparent mechanism for having advanced notice of emerging issues and related societal concerns in situations of normality and emergency or, in times of urgency, crisis. This includes an implicit dialogue between EU and national levels as well as with the public. For instance, picking up signals of concern from the public debate, or a consumer complaint line, for responding to early warning indicators, brings more consistency into the policy-making decisions and review process. However, questions loom up as to which extent should this be done, and how to organising it into a coherent structure and applying it. 'What is realistic for doing so?' In addition, risk assessors are supposed to take up new operational definitions like framing review, framing referral and framing terms of reference (ToR) in their repertory.

As stated in the SAFE FOODS paper, there is a descent obstacle. Specific attention must be drawn to the principle of non-delegation, as expressed in the 'Meroni' doctrine. It is still the dominant argumentation framework both in legal and political debates for restricting tendencies of functional decentralization in the European institutional structure. It appears that EU Member States have echoed in some way the developments in Brussels and Parma. This is especially relevant with regard to the mandating of tasks for risk assessors of EFSA and/or national food authorities/agencies (i.e. countervailing power and self-tasking) and the evaluation phase where the advice is handed in by risk assessors (i.e. timing and working on). As a result, the set up of real and harmonized boundary arrangements should also define concrete procedures for the involvement of national authorities/agencies to the work of EFSA. If a Member State requests a scientific opinion from its national authority/agency and the Commission does not consult EFSA, there is a risk of disputes and confusion, leading to divergence concerning the risk management measures. It is noted that the procedure for solving conflicts provided for in Regulation (EC) 178/2002, Article 30(4) has, so far, not properly functioned (e.g. GMOs, semicarbazides, TSEs[4]).

Henceforth, I would like to comment on some of the SAFE FOODS suggestions given for procedural reforms in risk analysis, with a specific focus on the element of risk assessment including the interface arrangements proposed. In doing so, I would like to start with a series of common concerns regarding current practice in risk assessment. Properly and realistically interpreted, the assessment of risk must be undertaken in an independent, objective and transparent manner based on the best available science. Above all, risk assessment is a scientifically based process of evaluating (putative) hazards and the likelihood of exposure to those hazards,

[4] Transmissible Spongiform Encephalopathy.

and then estimating the resulting impact on human, animal, plant or environmental health. In the end, risk characterization serves to bridge risk assessment and risk communication, allowing for the discussion of confidence and uncertainties in the analysis. It provides a scientific framework for understanding the impact of a wide variety of variables. Considering the strengths, science and scientific advice should only be performed by the well-known rationalists, but not merely from natural sciences (correct framing). This yields several logical questions, like: 'What are the factors that result in risk to public health?' 'How much harm could occur, and when could it occur (i.e. primary appraisal)?' 'To what extent can that harm be reduced by various intervention strategies (i.e. risk management options)?' and 'Which are the measures for individuals to keep the risk under their personal control (i.e. secondary appraisal)?' Interestingly, science is also a value judgment by using mode of expressions, such as: 'based on sound science', 'scientific evidence tells us', 'best practice', 'science based' (Mumpower & Stewart 1996; Hoppe 2007).

Despite a consensus on the central role of a science-based risk assessment, the usability for risk management and policy-making is facing problems. It is clear that the rational and 'science-based' approach implies a need for improvement. Firstly, science in itself is always characterized by uncertainty. Secondly, is it the condition of uncertainty and socio-political ambiguity, which is especially high for so-called new or emerging risks, where no sufficient historical data can serve as a benchmark for assessing the probabilities and adverse effects. Actually, new threats are emerging all the time. Thirdly, there is a major battleground over the concept and interpretation of the precautionary principle, the acceptance or otherwise of risk analysis, and the supposed requirement of science to prove safety. Traditionally, the precautionary principle should be invoked where the scientific evidence for safety is insufficient, inconclusive, or uncertain, or where preliminary scientific evaluation suggests that effects on safety may be unacceptable and/or inconsistent with the chosen level of protection. As such, it should be the realm of the policy-maker. However, this seems to be unrealistic and too limited an approach for risk management (de Hollander & Hanemaaijer 2003; Stirling 2007). Instead, as shown by the SAFE FOODS project, its adoption in the risk assessment stage may offer a new way for conditions where quantitative risk-based methods do not apply (i.e. uncertainty, ambiguity, and ignorance). Fourthly, it is becoming increasingly clear that the public feels scientific reasoning to be difficult to follow, that it embraces different values and is unclear to them, for example, what is regarded as 'uncertain', 'negligible' or within 'natural variability or boundaries'. Fifthly, even scientists are used to an uncertain world; a situation of socio-political ambiguity cannot be solved by a conventional, science-based risk assessment alone. Sixthly, how to make the advice and argument credible represents a dilemma. It depends on whether or not the reader can actually follow the steps taken by risk assessors, scientific committees or panels. Obviously, this solves the paradox between substantive and procedural transparency: objectification of lay perspectives and objectification of decision-making (Bal, Bijker, & Hendricks 2004). For the assessment, the reader (layman) should be able to think along with the rationalists (scientific expert). Seventhly, risk assessors are often impatient with those who oppose them, and forget to ensure that their opinions are explained clearly and simply to, for instance, policy makers, risk managers, industrialists, and the public at large. A failure to address the concerns of end-users has consequences for confidence and trust. People's risk perceptions determine what they characterize as a risk and how they will react to different hazards (Frewer et al. 2004). At present, not addressing the concerns is one of the causal factors associated with the decline in public confidence in risk assessment and, ultimately, in trust in risk management measures. Eighthly, it is difficult to implement into a science-based process the democratic right to be involved as non-experts. However, as aforementioned, it is important to address people's perceptions and their values. Otherwise the activities of risk assessors are likely to be considered as detached from society, and the public will distrust the motives of risk managers. Ninthly, the requested consideration of cultural and human values shifts the regulators' focus from risk only to a quite different task of balancing risks versus benefits in deciding on substances in foods or production technologies. The benefit assessment should be scientifically analysed with the same rigour as employed for the risks by implementing it as an independent stage. Finally, it is observed that there is the lack of shared definitions and practices across European borders in handling risk issues. Here, a rather striking picture turns up. The institutional structures and legislative frames of national food authorities/agencies in Member States appear to be as variable as a chameleon including expressions of divergent cultural attitudes (Vos & Wendler 2006).

So far, it can be concluded that the research of the SAFE FOODS project has covered all the aforementioned concerns and issues. For the risk assessment, the team has transferred them into innovations and promising suggestions for change. In brief, if the science-based risk assessment cannot provide non-ambiguous information, additional rules for deriving risk management strategies have been formulated. Apart from adverse effects, and depending on the respective risk types or conditions of threat, other technical assessments are being taken into account in order to select appropriate risk management options. Against this background, conclusions based on the framework proposed in the SAFE FOODS project should transfer easily to other representations of the risk analysis process, such as those presented by the United Nation's Food and Agriculture Organization (FAO), Codex Alimentarius, and the Organization for Economic Co-operation and Development (OECD), as well as those at national and EU levels, because the food safety governance activities represented by the SAFE FOODS framework show risk assessment as a distinct component driven by problem definition activities and feeding into decision-making activities via interface processes as a co-operative exercise.

However, there is still a long way to go in order to develop a systematic approach that is not seen as infringing upon the prevailing view on the delegation of powers ('Meroni' doctrine) or the full responsibility of the Commission or Member States governments for the conduct of risk management. But giving only very specific and limited powers to independent advisory bodies, such as EFSA and its national counterparts, contradicts the genius of Regulation (EC) 178/2002. Last, but not least, the General Food Law requests for a (more) pro-active attitude of risk management. This includes by definition self-tasking and unsolicited advice of risk assessors and, related to that, an independent problem definition (framing). Consequently, it will be carried out as a co-operative exercise between mainly risk assessors at EU and national levels and less by requests of risk managers. This will lead to developing own risk assessment strategies between EFSA and counterparts in Member States. Therefore, I disagree with a point uttered by critics of the SAFE FOODS team claiming this to be one of EFSA's shortcomings. On the contrary, it is a prerequisite and added-value, society asks to search for.

Nevertheless, the initiatives of the SAFE FOODS project should be encouraged, because from the analysis of the GM controversy and recent food safety crises (especially the BSE crisis) it can be concluded that European regulatory authorities have failed to take account of what drives public concerns (Frewer et al. 2004; Abels 2002). Current real life practice in food safety governance is not involving formal steps for shared understanding of the objectives at stake. Today, it possesses a quasi-exclusive focus on risks for human, animal and environmental health and generally convenes informal consultations at the discretion of civil servants (i.e. EC risk managers). Beside scientific expertise and integrity, as well as a precautionary measure concerning scientific uncertainty, the proposed reform of the risk analysis paradigm addresses both a constructive implicit and explicit dialogue between expert assessors and non-experts such as risk managers and other stakeholders. An explicit dialogue consists of balancing and assignment of trade-offs with affected target groups or the public at large (acceptability of distribution of risk, benefits and costs).

Among others, the choice of consultation tools depends on who needs to be consulted, and on available time and resources. The concept of proportionality of consultation to importance of the issue is emphasized here. This requires a ranking of the issue at stake in relation to other issues. This requirement has been worked out by SAFE FOODS as a participatory and acknowledged framing and screening step. Based on the concept of coping rationally with risks (de Hollander & Hanemaaijer 2003) the suggested 'General Framework for Precautionary and Inclusive Governance of Food Safety' can reach consensus on the problem and expediency. All these elements shape the question(s) to be answered by the specialists in the risk–benefit assessment stage. EU governments have maintained a policy on risks geared to equal protection of all members of the population. However, this has not always proved feasible in practice. For example, if there is no uncertainty or ambiguity under conditions of use, risks can be calculated as probability x effect (Kaplan & Garrick 1981) and safety standards are in place. However, the risk analysis concept does not consist exclusively of objectively science-based characteristics of food systems, but it is a social contract. Questions such as 'What about the degree to which the activity is voluntary?', or 'How fair are the joys and burdens distributed?', or 'To what extent is the situation manageable?' count, too. The framework in realising a transparent approach, according to the SAFE FOODS project, adds the process to select the most appropriate, efficient and proportionate strategy to assess the harmful as well as beneficial effects to human, animal and plant health. In this respect, the scientific risk–benefit analysis of food should be incorporated into Regulation (EC) 178/2002. The SAFE FOODS framework indicates also the level of participation required. In realising a transparent and more participative approach, therefore, the risk governance escalator of Klinke, Dreyer, Renn, Stirling, and VanZwanenberg (2006) and de Hollander and Hanemaaijer (2004) should be introduced in current protocols. In this way – from certain and unambiguous via complex to uncertain and ambiguous conditions – a matrix can be created on which the various types of objectives can be profiled into categories of necessary risk–benefit and socio-economic assessments, and rank the policy-making decisions requiring various numbers of participating actors (Renn 2004; Klinke & Renn 2002). A difficulty may be how to classify in advance the objectives where there is a high degree of ambiguity. The mechanisms for (possible) threats in emerging technologies are often unknown to experts and society, which leaves open many uncertain consequences that are dreaded by the population. Acquiring experience and confidence may also mean that subsequent innovations of, for instance, agri-biotechnology become more familiar to people. Thus, there is a moving benchmark in defining the objectives for regulatory actions. A key task for EFSA, DG SANCO, and Member States is the development of high quality interfaces, which are recognized as truly authoritative both within the EU and in the wider international arena. These communications in their day-to-day boundary work need to be based on sound normative principles involving Europe's regulatory framework. Such interfaces can be resource intensive, however. Therefore, careful consideration needs to be given to the question of how the current infrastructures available within the EU can be best utilized. The SAFE FOODS options for the institutional setting of the interface stages with the aim of rendering the information exchange more transparent and inclusive as well as of achieving better co-ordination, are considered in more detail below.

Overall, a participatory framing step will contribute to a transparent, consciously shared understanding of the objectives. Certainly, it may lead to plausible decisions between an intervention because of costs and, for instance, one based on equal protection of the whole population. In this way, subjective perception aspects are integrated into science-based risk assessments and definitively concluded by the risk assessor. This should lead to a consistent manner of differentiating types of risks including the perceived ones. SAFE FOODS has initiated discussions on the modalities of organising this expertise at EU level. However, I attach also importance to comparable efforts at national level in order to encourage a concerted and more systematically based transmission to Member States of requests sent to EFSA.

Political processes on the acceptance of technological innovations in food production show that scientific contributions will rarely be conclusive, given the uncertainties of 'scientific facts' (Funtovicz et al. 2000). It should be elaborated whether managerial or legislative action is required, for instance, in line with ethical values and distribution of the risk, the benefits and costs (fairness principle). SAFE FOODS acknowledges that science alone cannot solve uncertain and ambiguous phenomena that are associated with value disputes in society (Meyer et al. 2005). Its participatory ranking of community policy-making is proposed to prevent escalation and hardening of the conflict between proponents and opponents (see also Gaskell 2004). Once public concerns, for instance, about uncertainties of long-term biosafety are understood, they can, eventually, more effectively be introduced in the ranking of management options and measures. However, public participation and consultation in the risk analysis paradigm is still a very young concept (Dietrich & Schibeci 2003). It needs further developing in order to meet the demands of inclusive and accountable food safety governance (Renn 2004; Abels 2002). Flexible options such as public hearings, round tables, consultations, or other explicit dialogues, have preference to be selected. It is aimed at reaching consensus on equilibrium and fairness in decision options and legislative measures. The latter could be formulated in a pre-advisory Inter- and/or Extranet report presented to the audience of stakeholders and representatives of consumers (the public). It increases the acceptance of the risk–benefit assessment, actions, allows balancing pros and cons and possibly eliminates the need for relatively expensive options of governance. Such an assessment would need to gather new disciplines and fields of competences.

Indeed, there is a need for change. Especially, the key challenge facing food safety governance is to improve the implicit and explicit dialogue with the public. It is proposed that scientific values (technical objectives) and consumer values (social and economic or ethical concerns) are best settled in parallel, not consecutively, in the risk assessment stage. It is incumbent on specialists, industrialists and regulators to continue to integrate the public in emerging technologies and their respective risk–benefit and/or socio-economic assessment and ranking of decision options. However, the proposed participatory process of SAFE FOODS, a truly interdisciplinary governance approach, requires a solid knowledge of group interactions and incentives at EU and national levels, of technical, social and cultural competence, and memorable practical experience (i.e. anecdotal and systematic evidence). How to introduce the interface innovations of the SAFE FOODS project into current practices, without an overkill of participatory procedures, presents a formidable series of challenges in the future. Related to this, the improved Internet dialogue is highly praised as an example of inclusive and accountable means of policy-making (Abels 2002). A specific Inter- and/or Extranet Forum would allow for a broad input from a wide variety of interest groups, civil society organizations and the wider public, without the burden to pre-select. Its value is only materialized if the results are used in both the framing and the evaluation stage of the risk analysis paradigm. Overall, dialogues need to be held in a transparent and flexible manner in order to raise awareness in society that people have the ability to let their voice be heard.

To conclude, effective documented institutional mechanisms for achieving deliberation and participation are yet rarely implemented or evaluated. Thus, it is still questioned whether the SAFE FOODS framework could be a 'best practice' of such safety governance model. Discursive procedures will increment by options such as convening meetings with interdisciplinary expert groups to discuss specific objectives pertaining to risk–benefit and/or socio-economic assessments of factual consequences of each option to mitigate the risk. Organising more flexible ad-hoc meetings of representatives of (organized) stakeholders to identify their values, concerns, criteria and incentives will increasingly become important in the Europe's decision-making process. Thereto, more particular procedures for co-operation between EFSA and its counterparts in Member States should be established. Facilitating the exchange of views between the public, experts and regulators possibly through the Inter- and/or Extranet (i.e. flexible consultations) seems to provide the most suitable platform for this purpose, and evaluates and ranks the decision options for risk management given the different world views and interests of stakeholders. Given these directions, taking into account the viewpoints of different stakeholders and establishing independent assessments is a major challenge in safety governance of technologies associated with uncertain and ambiguous risks such as GMO-, nano- and cloning-technologies.

As stated before, regarding the options for the framing and evaluation interface, it depends on whether the reader is able to think along with the rationalists (Bal et al. 2004). Therefore, a discussion of the question put to EFSA or a national risk assessment body in an Inter- and/or Extranet Forum, which could be a mandatory consultative procedure in, respectively, the setting of EFSA and Member States tasks or the conduct of evaluation, might be the preferred option. Moreover, it should be recognized that, at present, food safety systems and institutional responsibilities for risk assessment differ among the twenty-seven European countries. As described in the Advisory Forum document 'Strategy for cooperation and networking between the EU Member States and EFSA' (2006), EFSA will give highest priority and appropriate resources to developing the practical infrastructure necessary for the greater involvement of the Member States. In particular, EFSA's formulation and delineation of Focal Points and the Extranet, which facilitate true exchanges of scientific and communication information, develops into a fully functioning and active tool. As such, it can be postulated that simply harmonising risk assessment methodologies and consultation approaches in the EU would already contribute to an enhanced level of confidence and ultimately the mutual recognition of scientific opinions across Europe. Such an operating way would also enable an efficient use of resources and competences existing in all authorities/agencies of Member States. It is in the remit of EFSA's Scientific Committee to define these harmonized approaches. Prior information of national risk assessment bodies of EFSA's forthcoming work and progress would be needed, too. Last, but not least, pro-active attitudes of food safety governance demand a degree of independent self-tasking at the side of risk assessors; however, at present, policy-makers and risk managers are showing signs of being over-anxious not to commit themselves (i.e. 'Meroni' doctrine).

References

Abels, G. (2002). Experts, citizens, and eurocrats – towards a policy shift in the governance of biopolitics in the EU. *European Integration Online Papers*, 6(19). http://eiop.or.at/eiop/texte/2002–019a.htm. Accessed 15 January 2008.

Bal, R., Bijker, W. E., & Hendriks, R. (2004). Democratisation of scientific advice. *British Medical Journal*, *329*, 1339–1341.

Codex Alimentarius (2007). *Working Principles for Risk Analysis for Food Safety for Application by Governments*. CAC/GL 62/2007.
Commission of the European Communities (2000a). *Communication from the Commission on the Precautionary Principle*, COM 1 final, 2 February 2000. Brussels.
Commission of the European Communities (2000b). *First Report of the Scientific Steering Committee's Working Group on the Harmonisation of Risk Assessment Procedures*. Brussels: Health and Consumer Protection Directorate-General.
Commission of the European Communities (2001). *European Governance. A White Paper*, COM 428 final, 25 July 2001. Brussels.
Commission of the European Communities (2003). *The Future of Risk Assessment in the European Union*. Europe: Scientific Steering Committee.
Commission's DG SANCO (2005). *Maximising the Contribution of Science to European Health and Safety*. Brussels: DG SANCO.
de Hollander, A. E. M. & Hanemaaijer, A. H. (Eds.) (2003). *Coping Rationally with Risks*. RIVM rapport 251701047 (pp 1–52). Bilthoven, The Netherlands: National Institute for Public Health and the Environment (RIVM).
Dietrich, H., & Schibeci, R. (2003). Beyond public perceptions of gene technology: community participation in public policy in Australia. *Public Understanding of Science*, *12*, 381–401.
EFSA's Scientific Committee. (2006). Transparency in Risk Assessment carried out by EFSA. *EFSA Journal*, *353*, 1–16.
Frewer, L. J., Lassen, J., Kettlitz, B., Scholderer, J., Beekman, V., & Berdal, K. G. (2004). Societal aspects of genetically modified foods. *Food and Chemical Toxicology*, *42*, 1191–1193.
Funtovicz, S., Shepherd, I., Wilkinson, D., & Ravetz, J. (2000). Science and governance in the European Union: a contribution to the debate. *Science and Public Policy*, *27(5)*, 327–336.
Gaskell, G. (2004). Science policy and society: the British debate over GM agriculture. *Current Opinion in Biotechnology*, *15*, 241–245.
Hoppe, R. (2005). Rethinking the science-policy nexus: from knowledge utilization and science technology studies to types of boundary arrangements. *Poiesis and Praxis*: *International Journal of Technology Assessment and Ethics of Science*, *3(3)*, 199–215.
Hoppe, R. (2007). *Scientic Expertise and the Policy Process: Boundary Workers' Perspectives, Paper for Interpretive Policy Analysis Conference*. Amsterdam: Free University Amsterdam.
Kaplan, S., & Garrick, B. J. (1981). On the quantitative definition of risk. *Risk Analysis*, 1, 11–27.
Klinke, A., & Renn, O. (2002). A new approach to risk evaluation and management: risk- based, precaution-based and discourse-based strategies. *Risk Analysis*, *22*, 1071–1094.
Klinke, A., Dreyer, M., Renn, O., Stirling, A., & van Zwanenberg, P. (2006). Precautionary risk regulation in European governance. *Journal of Risk Research*, *4(9)*, 373–392.
Levidow, L., Carr, S., & Wield, D. (2005). European Union regulation of agri-biotechnology: precautionary links between science, expertise and policy. *Science and Policy*, *27*, 261–276.
Meyer, G., Paldam Folker, A., Bagger Jørgenson, R., Krayer von Krauss, M., Sandøe, P., & Tveit, G. (2005). The factualisation of uncertainty: Risks, politics, and genetically modified crops–a case of rape. *Agriculture and Human Values*, *22*, 235–242.
Mumpower, J. L. & Stewart, T. R. (1996). Expert judgement and expert disagreement. *Thinking and Reasoning*, *2(2–3)*, 191–212.
Renn, O. (2004). *Deliberative Approaches to Manage Systemic Risks*, Presentation at the Euroscience Open Forum (ESOF). Stockholm: ESOF. http://www.esof2004.org/pdf_ppt/session_material/ortwin_renn_2.ppt. Accessed 15 January 2008.
Stirling, A. (2007). Risk, precaution and science: towards a more constructive policy debate. *EMBO reports*, *8(4)*, 309–315.
Vos, E., & Wendler, F. (Eds.). (2006). *Food Safety Regulation in Europe*: *A Comparative Institutional Analysis (Series Ius Commune)*. Antwerp: Intersentia.

12.3 A Consumers' Association's Perspective on the Governance Framework

Commentary from Sue Davies

12.3.1 Introduction

The approach to controlling food safety risks is always controversial. This has been especially the case over the past two decades which have seen European consumers exposed to a range of food safety risks of differing magnitude and complexity.

As the authors highlight, the inability of policy-makers to deal effectively with the risks posed by Bovine Spongiform Encephalopathy (BSE) in particular prompted a review of the institutions that have responsibility for dealing with food safety within many member states and at European level, most notably with the creation of a European Food Safety Authority (EFSA).

For many years, as the far-reaching consequences of the BSE crisis have become apparent in public health, social and economic terms, food safety has been at the top of the political agenda. Issues such as the use of hormones in beef and genetically modified (GM) foods have also highlighted the way that food risks are global, and involve a complex interaction between scientific considerations and broader social and economic considerations which may often conflict.

While much has been learned from the way that these problems were dealt with, particularly a greater acknowledgement that decisions will have to be made when there may be a great deal of scientific uncertainty, many challenges still remain.

It is difficult to anticipate what scares may be on the horizon, but issues currently on the agenda, including the use of animal cloning for food production and a broad range of potential applications that rely on nanotechnologies, are again bringing into focus the need to take a precautionary yet proportionate approach to dealing with food risks that are about much more than scientific risk assessment.

It is therefore a crucial time to be reviewing the framework for dealing with food safety risks. This is not merely because there will always be new and complex issues to consider, but also because it continues to be necessary to ensure that the stark lessons and principles that were top of mind for politicians and policy makers in the wake of BSE, are not forgotten and subsumed by too short-term a focus and the current political imperative to reduce the burden of regulation on industry.

12.3.2 Governance Challenges

The authors identify five main governance challenges which deserve further attention: the organization and relationship between risk assessment and risk management; dealing with scientific uncertainty; the handling of highly controversial food safety issues; establishing transparency during the entire food safety governance process; and the provision of effective and legitimate mechanisms for stakeholder and public engagement. They recognize that recent EU level reforms have addressed them to some extent, but these have not gone far enough.

Three further challenges, which are explored to some extent in the model, should also be given greater prominence. As well as the interface between risk assessment and risk management, the interface with risk communication is also important. This includes ensuring greater efforts to have a broader understanding of what risk communication means and why it is still, all too often, seen as a top-down process, rather than a two-way exchange. The mechanisms proposed for establishing stakeholder dialogue at key stages in the decision-making process are clearly aimed at addressing this. This aspect is particularly relevant given that risk communication is a responsibility shared by the European Commission, Member States, and by EFSA.

The second additional challenge is the need to be more pro-active in identifying emerging risks. While there has been a lot of attention given to how horizon scanning can be carried out, and EFSA has established a unit to specifically focus on new and emerging risks, it remains challenging in the complex global environment with long and integrated supply chains, as highlighted by the illegal contamination of foods with the dye, Sudan I. When risks are identified, it is also essential that the principles that are established as part of a more precautionary and integrated framework do not result in unnecessary delays that put more consumers at risk, even if they have the best long-term intentions.

The third additional challenge which needs to be given specific prominence is the necessity to better integrate 'other legitimate factors'. This goes to the heart of the changes that the authors suggest, giving much greater prominence to the evaluation of these factors alongside the scientific assessment. The role of 'other legitimate factors' has been recognized in EU legislation including the General Food Law Regulation[5] and the GM food and feed regulations[6]. They are also explicitly referred to in the Codex Working Principles for Risk Analysis for Food Safety for Application by Governments (Codex Alimentarius Commission 2007). In a UK context, the Food Standards Agency has responsibility for protecting the health of consumers, but also for protecting other consumer interests in relation to food (UK Food Standards Act 1999), although this responsibility has remained poorly defined. However, it remains unclear how much weight will be given to these 'other factors' in practice, particularly if they are at odds with the scientific risk assessment.

The objectives of the model are therefore the right ones: offering a truly interdisciplinary governance approach; more consistent application of the precautionary principle and improving the co-ordination between risk assessment and risk management, placing much greater emphasis on the framing and evaluation stages.

Underlying all of these considerations is the relationship between the action that is possible by Member States, by the European Union institutions and how these

[5] General Food Law Regulation (EC) 178/2002.

[6] Regulation (EC) 1829/2003 on genetically modified food and feed, Article 7.

relate to World Trade Organization (WTO) commitments and, therefore, given its special status under the WTO agreements, risk analysis principles as recognized by the Codex Alimentarius Commission. This relationship is fundamental to sustaining any food safety risk management decisions.

12.3.3 Progress to Date

The authors acknowledge the substantial institutional reform that has taken place following the succession of food safety scares in the 1990s and a breakdown in confidence in the institutions and mechanisms for handling food safety. At European level this led to a strengthening of the Directorate General for Health and Consumer Protection (DG SANCO) with responsibility for risk management and communication and the establishment of a European Food Safety Authority (EFSA) with responsibility for risk assessment and also risk communication. In conjunction with this, a White Paper on Food Safety (Commission of the European Communities (CEC) 2000) set out a range of legislation that was to be reviewed or newly introduced in order to enhance consumer protection. The precautionary principle was enshrined within the General Food Law Regulation in recognition of the disastrous consequences of failing to apply precaution across risk analysis.

The authors summarize the responses seen, into three main themes: procedural and structural mechanisms designed to assure a stricter separation of risk assessment and risk management and, therefore, enhance the independence of risk assessment; the growing attention to and communication about scientific uncertainties and advancement of the democratic quality of the governance process through greater public engagement. Greater openness, transparency and public involvement in decision-making remain the fundamental cornerstones of a more effective way of dealing with food safety risks. There have been positive steps taken, but these themes have been implemented to a greater or lesser extent depending on the issue, the institution and the responsible policy division within the institution. It is therefore appropriate and important to review how these principles can be applied more systematically.

12.3.4 The General Framework

Risk analysis is generally seen in three distinct but interactive stages: *risk assessment, risk management* and *risk communication*. This is the approach that has been developed following successive Food and Agriculture Organization (FAO) and World Health Organization (WHO) consultations. After years of debate, Codex finally adopted risk analysis principles directed to governments in July 2007 that recommend such an approach, with risk assessment policy explicitly recognized as a component of risk management, but cutting across the risk assessment and risk management responsibilities.

The General Framework that is proposed includes four different, iterative stages: *framing, assessment, evaluation* and *management*. It is described as an 'open, cyclical, iterative and interlinked process'. The two new stages of framing and evaluation are a significant improvement, but as it has taken over 10 years to reach agreement on a common language at international level it, therefore, seems preferable to see the underlying, important concepts that the authors are proposing integrated more effectively within the internationally adopted framework, for example as components of risk assessment policy and risk management.

12.3.4.1 Framing

One of the main strengths of the General Framework is that it gives much more specific acknowledgement of the 'framing' part of any decision about how to manage food safety risks. Although this is implicit from the point that an issue is first identified either by the European Commission, EFSA, or by Member States, it still appears to receive too little attention in the way that the current framework operates. The outcome relies on the right questions being asked from the outset, and from them being addressed by the right bodies. More explicit consideration of how issues are framed at the very beginning could lead to more robust and successful (i.e. socially acceptable) outcomes.

The three stages of the 'framing process', review, referral and terms of reference, although often implicit, are currently not being adequately and transparently considered. The 'review stage' is where the broader policy and legislative framework is established. The 'referral' stage of framing is where a particular issue is seen as being forwarded to EFSA, while also being placed for review by stakeholder would formalize a process for commenting on the draft terms of reference.

The Framework would help to ensure that the discussions around the framing or establishment of the terms of reference are more inclusive and take account of a broader range of stakeholder views. However, the Framework could also benefit by setting out more explicitly the stage before the terms of reference are sent to EFSA. Although in most cases, EFSA will be the responsible agency, it may also be necessary to involve others. It is therefore essential that this 'framing' stage is sufficiently broad at the very outset. A recent example of how this may work was seen when the European Commission requested an opinion from EFSA on the food safety, animal health and animal welfare aspects of cloning in food production, while also requesting an opinion on the ethical aspects from the European Group on Ethics in Science and New Technologies (EGE).

Two new bodies are proposed which would contribute to the framing process as well as to the evaluation: an Internet Forum and an Interface Advisory Committee. These are, in principle, positive proposals but they raise several questions that need to be addressed further. The Internet Forum would enable debate among a broad range of stakeholders and experts. However, as with the creation of any new body of this kind, there would need to be clear rules of procedure to ensure that, in an effort to be more inclusive, a more exclusive process was not instead introduced.

It would, therefore, be essential to ensure that the breadth of interests across the food chain were represented on the Forum, including consumer organizations who may be more limited in terms of having the resources to actively participate. It would also need to be ensured that the process was transparent, including clear declaration of interests and that it did not merely become a limited forum for debate among a group of 'usual suspects' with the most time to contribute, but not necessarily the most relevant expertise or representativeness.

Another new body, the *Interface Advisory Committee*, would have responsibility for defining the detailed terms of reference. This would build on the current exchanges between the European Commission and EFSA by involving stakeholders. While again, a potentially positive development in principle, this raises several issues and unanswered questions that would need to be addressed.

It is suggested that the Interface Advisory Committee would not be expected to deal with all cases of risk governance, but only address those cases considered to be particularly problematic or requiring further discussion between risk assessors, risk managers and stakeholders. If the role of this Committee is to help frame the issue by establishing the terms of reference, who has responsibility for the initial framing and determines whether or not this is an issue that requires the involvement of the Advisory Committee?

It is proposed that the Interface Advisory Committee would work in a flexible setting, with its composition depending on the case in question with a core group of permanent members. It is proposed that the Commission would appoint these core members including two to four risk assessors, risk managers, and stakeholder representatives. While it is clearly beneficial to keep such a group manageable in size, it would be difficult for two to four individuals to be appointed as representatives of all stakeholders. The EFSA Stakeholder Consultative Platform, for example, includes 24 organizations from across the food supply chain. While clearly it may be possible to narrow this down for the purpose suggested, even with such a comparatively large membership, EFSA is regularly criticized for excluding groups that feel they should be represented. There is, therefore, a danger that the group would allow two to four stakeholders to have a significant amount of influence. It would need to be more broad-based to be effective. This could then raise the issue of whether such a role could be dealt with through some modification or transformation of other existing and relatively recently introduced stakeholder fora, the Advisory Group on the Food Chain and Animal Health and the EFSA Stakeholder Consultative Platform which may involve the same stakeholder groups and even the same individuals.

Transparent methods of working would obviously be critical if the Committee were to have any credibility, including open meetings. Clear rules for handling of requests for commercial confidentiality would be essential. A further question it raises is the implications for resources. The Interface Advisory Committee would initially be of interest to a wide range of organizations, but how could it be ensured that an effective and representative range of interests were able to fully participate in the discussions on a regular basis and that a balance could be maintained in the longer-term?

12.3.4.2 Assessment

The process of assessment includes a more explicit 'screening' phase than is currently the case. This categorizes issues as either serious, uncertain or ambiguous. This is seen as corresponding with Codex's and WTO's 'preliminary risk assessment' stage.

This screening stage is useful in that it requires explicit consideration of the nature of the risk and determines how much attention it will subsequently receive. If the issue is considered to be 'certainly and unambiguously serious', preventive measures will be considered. If there is scientific uncertainty or ignorance, a useful and important distinction, the threat is assigned to a precautionary assessment. If it is considered to be 'socio-politically ambiguous' it will undergo a process of concern assessment. If the issue is not considered to be either serious, uncertain or ambiguous, it will undergo a conventional risk assessment.

While it is useful to consider the nature of different risks, or threats, in this way, there appears to be a danger that a risk may be wrongly assigned from the outset, potentially limiting the scope of the assessment. A lot of responsibility is placed on EFSA staff for assigning the nature of the assessment, based on a proposed set of criteria. However, these also include aspects that may lie outside EFSA's current remit, such as the 'social criterion' ('whether there are signs of adverse effects in terms of social justice in the distribution of threat or in terms of manifest political mobilization on the part of particular public constituencies'). It is proposed that a 'concern assessment' unit would be established within EFSA to help with this and broaden EFSA's expertise into areas of social science.

In practice, issues may cut across all of these categorizations. The terminology also suggests that some assessments need not be precautionary, whereas precaution needs to be implicit across all assessments. Similarly, the Framework appears to narrow down the scope for consideration of socio-economic factors. For example, it is suggested that if a conventional assessment reveals risks to be low in magnitude, it would not be effective or proportionate to include detailed assessment of socio-economic factors. The assumption is that *as the magnitudes of risk are recognized to increase, there will be a corresponding necessity to provide subsequent evaluation and management stages with information concerning the nature and scale of any socio-economic benefits or justifications for the toleration of what might otherwise be seen as relatively high levels of risk.* This could, however, apply equally where the risk is considered to be low but there may be public concern because of the perceived imbalance between the risks and benefits, the nature of the risk, or if any groups, such as children, are particularly likely to be affected.

The approach to assessment set out in the Framework is, therefore, positive in that it would ensure that some considerations, assumptions and categorizations that are currently made *implicitly*, would be made *explicitly*. However, it is important that the Framework does not inadvertently lead to an unnecessary narrowing of the approach to assessment too soon and based on poorly informed assumptions. The importance of inter-linkages is stressed, but clear processes that allow flexibility and ongoing review without unnecessary delay will be essential.

It is also essential that there is collaboration across EFSA so that all relevant Panels and expertise are drawn upon, for example, nutritional considerations relevant to toxicological issues. There should also be a mechanism for interaction with the relevant authorities that have practical experience of the implementation and enforcement of measures that may be considered as part of the assessment, including Member States (for example through EFSA's Advisory Forum), and the European Commission through its Food and Veterinary Office.

More generally, the Framework should give greater emphasis to the importance of ensuring the transparency of the assessment bodies. While EFSA has, for example, taken steps to open up its scientific panels, including clearer rules for declaration of interests, consultation on draft opinions and greater interaction between the Scientific Committee and Panels and the Stakeholder Consultative Platform, further steps are needed. These include, for example, holding open meetings of the scientific committees and appointing public interest representatives to the Committee and Panels. This has been shown to work well in the UK, improving transparency and enhancing the robustness of the opinions.

12.3.4.3 Evaluation and Management

The Framework includes 'evaluation' as a separate stage in the risk analysis process. Conventionally, it is included within risk management. This is a very useful distinction because, as already emphasized, it allows for more explicit consideration of the 'other legitimate factors' – the broader social and ethical aspects that will influence the approach to risk management, but are not always fully considered or clearly communicated when they are taken into account.

The opening up of this 'evaluation' stage should help to ensure that there is a transparent consideration of all of the issues. In some cases, for example, the decision may be dominated by the assessment, as the issue is quite straightforward. In other cases, the social and ethical considerations may far outweigh the scientific assessment, requiring risk managers to act on this basis. Clarity and transparency about how such decisions are reached is essential.

The proposal to place the evaluation step after the assessment stage could, however, be too limiting and lead to unnecessary delays. The evaluation stage could be taking place alongside the assessment after the screening. The reason for placing it after the assessment given by the authors is to enable insights from the assessment exercise to be summarized and deliberated. While it may be necessary to review these aspects together, waiting until the assessment is complete before the evaluation begins appears to unnecessarily hold up the process when the two stages will generally involve different considerations and ultimately it will be for the risk managers to weigh the two together in consultation with all interested parties, including stakeholders.

It is proposed that stakeholder involvement in the evaluation stage would be formalized through the Interface Advisory Committee. This Committee is seen as having an important function with regard to advising on what is acceptable or tolerable and it is suggested that they would present advice to the Commission with regard to evaluation decisions, or alternatively stakeholders would be involved through the Internet Forum. Concerns about the inclusiveness and representativeness of the Interface Advisory Committee have already been raised in the context of framing. A great deal of influence is potentially being placed in the hands of a small number of stakeholders and individuals who are also likely to have difficulty dealing with a very heavy workload.

It is very unlikely that any agreement could ever be reached across the diverse range of stakeholders that would need to be included on the Committee. The advice is therefore likely to set out a range of views depending on the interests of the stakeholders. Risk assessors and risk managers will also be represented on the Interface Committee. Consideration could also be given to developing a more independent and transparent approach to providing advice on the evaluation alongside an effective mechanism for stakeholder engagement. Relying solely on the Interface Committee could be too limiting and may not provide the level of breadth of advice that would be needed. It may be better suited to reviewing expert advice provided on social and economic aspects that has been sought alongside the assessment stage. The Framework does recognize that where a topic raises strong controversy and evaluation is highly ambiguous, a fully fledged participation process might be appropriate such as use of stakeholder roundtables, citizen forums, jurors or consensus conferences. These mechanisms are very important and it would be useful to give them greater prominence. In any case, it will be necessary for the Evaluation to draw on the available research and expertise that is relevant to the issue under consideration.

The ultimate decision about how to balance the results of the assessment and the evaluation (e.g. the distinction described between an intolerable, tolerable or acceptable decision) should also be the responsibility of risk managers, rather than the Interface Committee.

The ultimate risk management is described as having six stages: identification, assessment, evaluation and selection of possible management measures followed by their implementation and management. As described, it is essential that these stages are interlinked so that experience gained from monitoring of the measures is reviewed. As with the assessment phase, the Framework categorizes the possible management approaches as prevention, precaution-based, concern-based and risk-based. Again, the concern here is that it must be ensured that this does not unnecessarily limit the approach by limiting the nature of the measures that are considered at the outset. The management approaches should, therefore, be seen as possible tools that can be used, rather than a more rigid template.

As with the entire framework, there must be transparency about how risk management decisions are reached. It would need to be ensured that risk managers involved stakeholders in the deliberation and evaluation of the most appropriate measures to adopt and that this was not seen as being the responsibility of the 'evaluation' phase alone.

12.3.5 Conclusion

The proposed General Framework for the Precautionary and Inclusive Governance of Food Safety makes a very positive contribution towards establishing a more robust, precautionary, inclusive and transparent approach to dealing with food risks.

It makes explicit some of the stages and assumptions that are often embedded within risk assessment and risk management and most significantly, gives much greater prominence to the initial 'framing' of the issue and the subsequent 'evaluation' allowing for a fuller and more open consideration of the socio-economic considerations that inevitably have a bearing on the final risk management decision.

Further discussion is now needed on how the types of new bodies and mechanisms that have been suggested to enhance stakeholder involvement, such as the Internet Forum and Interface Advisory Committee, could effectively work in practice and not inadvertently make the risk analysis process more closed and exclusive. Additional consideration is also needed as to how these very important concepts can be integrated within the recently agreed framework for risk analysis that Codex has adopted, gaining international acceptance.

References

Codex Alimentarius Commission (2007). *Working Principles for Risk Analysis for Food Safety for Application by Governments*. CAC/GL 62–2007.

Commission of the European Communities (CEC) (2000). *White Paper on Food Safety*, COM (1999) 719 final, 12 January 2000, Brussels.

12.4 An Industry Perspective on the Governance Framework

Commentary from Ruth Rawling

12.4.1 Introduction

I was invited to participate in the discussions on the food safety governance framework developed within the fifth subproject of the SAFE FOODS project (so-called work package 5, hereinafter referred to as WP5) 'A General Framework for the Precautionary and Inclusive Governance of Food Safety in Europe' in my capacity as Chair of the Food and Feed Safety Section of COCERAL, the European Association of National Associations for the trading of grains and other raw materials for food and feed. I bear full responsibility for the remarks below which I write in a personal capacity, but I should make it clear that my comments have greatly benefited from discussion with other members of the COCERAL executive and secretariat.

In my capacity as head of Corporate Affairs in Europe for Cargill I also have links with other industry associations such as CIAA, the European Association of food manufacturers in which my company has direct membership, and FEFAC, the European Association of national associations of animal feed manufacturers. Discussions with representatives and members of these organizations, on what industry is seeking from food safety governance were also helpful in clarifying my own thinking. After my participation in this project had finished I became a member of the European Commission Directorate General SANCO's new stakeholder dialogue group.

Food and feed safety is a subject that Cargill takes extremely seriously and we had been veterans of using the HACCP[7] system in our food and feed ingredient manufacturing plants well before legislation required it. We also have a history of sharing innovations in food safety with our industry colleagues – such as refrigerated ocean-going transport of orange juice or steam pasteurization of carcass surfaces in meat processing plants.

From this perspective, I was honoured to be invited to attend the WP5 sessions. I was present at the meeting in Schloss Haigerloch in September 2006, where we were first made aware of the SAFE FOODS's WP5 work and I subsequently made a presentation representing an industry perspective at the workshop in May 2007 in Brussels. What follows is a slightly expanded version of that presentation.

12.4.2 What Does Industry Look for in Food Safety Governance?

Let me first set out some principles, before I come to the specifics of the food safety situation in Europe. These principles apply to a system of governance anywhere in the world. Fundamentally, what industry needs is a system that works, that is risk based, that produces good, manageable laws in which people have confidence and that genuinely reduces food safety risks in the system. The best laws are those that incite the right behaviour in handling food and food ingredients. Food scares and unsafe food on the market are of no help to industry trying to go about its daily business in the food sector. Moreover, if confidence drains away from a sector it becomes harder to recruit talented people to it, and yet such people are essential in building an innovative, safe food system for tomorrow. Loss of confidence can produce a vicious circle which spirals downwards and makes it ever more difficult to recover that confidence.

12.4.3 Six Principles for Food Safety Governance

Here are six principles that provide the foundation for a good system.

Firstly, we look for a system that takes as its starting point scientific rigour and independence of scientific assessment, freed from other influences. The assessment should be focussed on risks that might prevent a food being safe. Science-based risk assessment is not the only factor in a system of governance, but it is the necessary

[7]Hazard Analysis and Critical Control Points.

condition on which any system must be built. Sooner or later, if this is not the foundation of food safety regulation, the regulation will be shown to be flawed and need to be replaced. The real test of regulation is whether it helps improve food safety in the system – and that can only be based on sound science and its assessment by experts qualified through training and experience. Of course, some scientific areas are much more uncertain than others and that is where application of these principles is particularly needed.

Secondly, we look for transparency in decision-making, so that it is clear where decision-making authority lies. Clarity of accountability would be another way of expressing this thought.

Thirdly, we look for timeliness in decision-making, which is often linked closely to resources and available expertise. For the food industry this becomes important not only when doubts are raised about the safety of a food and speed of reaction is important but also in relation to the approval of new foods and food ingredients. Timeliness of approvals directly impacts the rate of innovation.

Fourthly, we look for good communication that provides clear messages on what is safe and not safe, so that everyone can understand what the science is telling us. Good communication should also let us know where things are in the process, for example when we are seeking authorization for new foods.

Fifthly, we look for governance that is able to prioritize and distinguish the important from the urgent.

Sixthly, we look for governance, particularly risk management, which can draft good laws that are capable of being implemented and reinforce good behaviour in the system. Such laws need to take account of the political culture in which they belong.

12.4.4 Applying the Principles to Food Safety Governance in Europe

Let me now take these principles and apply them to the system of food safety governance in Europe.

It is clear from such a set of principles why industry welcomed the establishment of the European Food Safety Authority, EFSA, as the body in charge of scientific risk assessment for food safety in Europe. One body coordinating the scientific expertise available throughout the Member States would both leverage the resources of the EU to the benefit of all its citizens and contribute to reducing risks of diverging approaches to food safety issues. Together with EFSA, the General Food Law of 2002 confirmed and reinforced the HACCP ('Hazard Analysis and Critical Control Points') approach to risk as the core of food safety within industry, and established clear accountabilities through the entire food and feed chain for operators to manage food safety of their own products. For some in industry this was burdensome, but for any company already trying to meet high food safety standards this was a major step forward.

As with the establishment of any new body, however, there will always be teething troubles to be worked through and regulations that need tweaking to make things work well in practice.

The researchers of the WP5 team are to be congratulated in taking a hard look at the system we now have with the Commission, the Member States, and EFSA.

12.4.5 WP5 Diagnosis

The diagnosis of the WP5 team of the outstanding challenges and needs in the new system is sound, focussing in particular, on

- The interface between the risk manager (the Commission with the Member States) and the risk assessor – EFSA – and whether the right questions are being asked by the right people to enable focussed work to be done
- The problem of scientific uncertainty and whether this requires a more formalized evaluation of the kind of risk assessment done and the options open to risk managers
- The question of how to deal with societal concerns
- Transparency – clarity around how the precautionary principle is being applied

These concerns are real ones and deserve thorough debate.

The WP5 team would like to formalize the process for framing issues between the risk manager and risk assessor, structure the assessment process, and formalize the evaluation of the risk assessment before the management step.

In our discussions in Haigerloch it became clear that at the core of much of the discussion was the third point listed above, namely the question of how societal concerns were being dealt with. There was general consensus that more needed to be done here for good and effective implementation of laws and even the establishment of better laws. Where societal concerns were not taken into account, laws could lack legitimacy because they remained controversial.

The WP5 researchers proposed adding a capability within EFSA to do a thorough risk assessment of societal concerns for any scientific issue that an initial discussion would show was likely to raise such concerns. They further split the risk assessment process in EFSA into 'risk based', 'precautionary' and 'prevention'. They proposed that clear decisions were needed up front, when EFSA tackled an issue about which risk assessment process was needed.

12.4.6 Industry's Perspective

While the assessment of the WP5 team of needs and gaps in the current system has value, their proposed solutions to these gaps through structural reforms deserve to be looked at with a more critical eye.

The concepts for the stages of governance have merit, although ultimately it is for the risk managers to decide how to take account of all the risk assessment they have asked for. There is a limit to how far the step of evaluation of risk assessment can do management's job for it. The concept of paying more attention to the framing stage by risk managers and risk assessors does sound sensible so that there is alignment between them going forward; similarly it seems appropriate that risk assessors and risk managers should together evaluate the risk assessment. It is for the managers and assessors to decide the best way to do that: a formal interface committee does not seem like the best idea. It is more a question of having a suitable process that needs to be followed than needing to create a new structure.

From industry's perspective, the only real innovation being proposed in risk assessment for EFSA is that of societal concerns or the 'concern assessment'. The other routes are all based on scientific risk assessment. Granted, EFSA would need to make clear, each time, the amount of scientific uncertainty in the risk assessment – but this is something that EFSA is already doing reasonably well, although it can continue to work on improving its communication. It would be concerning if the word 'scientific risk assessment' was not used for the three stages which are all science based – namely what the WP5 researchers call 'risk-based approach', 'precaution-based approach', and a 'prevention-based approach'. All of these approaches have to rely on the assessment of available science by experts qualified by training and experience. The differences relate only to context of the question and amount of science available to address the issue. We would view the minimizing of a risk-based approach to one option – the one where risk is clearly quantifiable – as potentially undermining the fundamentals of the approach on which EFSA is working. Moreover, risk assessment uses precaution even when data is available. To ascribe precaution to one approach therefore seems to undermine the precaution present in any risk assessment. In addition, if we think for a moment about EFSA's interaction with other food safety agencies around the world, a different use of the concept of risk assessment by EFSA compared to its peers seems likely to lead to confusion. On this point, therefore, we think the WP5 approach has the potential to confuse rather than clarify: there are only two concepts of risk assessment that are being looked at, not four: one is science-based risk and the other is societal-based risk.

There is clear value in the idea of a risk assessment of societal concerns on controversial topics. Such an assessment can better frame the options for risk management. As the issue was debated, however, it seemed that the risk manager, i.e. the Commission and the Member States, already have the power to request such a risk assessment should they want it. There is already a body set up under EU Law to look at ethical issues related to controversial topics, not exclusively for food topics, the European Group on Ethics in Science and New Technologies. Current law means that the Commission and the Council can request advice from this body on ethics – which seems to us to be one aspect of societal concerns. We saw no reason why a similar body could not be set up to look at broader societal concerns and give advice; or why a panel of experts could not be on hand to advise on this issue and how it might affect the framing of laws; or why stakeholder consultation could not be used to greater effect to get at societal concerns.

The one point on which we could not agree with the researchers was the idea that such a body should sit within EFSA. There are three main reasons for this.

12.4.7 'Societal Concerns' Do not Belong in EFSA

Firstly, we think societal concerns are different from the physical sciences. While science is international, societal concerns differ from country to country and are essentially cultural. If EFSA is going to establish its reputation as the premier source for scientific expertise on food and feed in Europe, by definition it will also need to have an international reputation and be able to work well across boundaries when Europe is short of particular expertise. Moreover, given the size of the EU market within the global food market, we would expect EFSA to contribute its expertise to the ongoing harmonization of global food safety standards, which will facilitate trade and global supply chains. We feel that including societal concerns in EFSA's remit would be a distraction in building that scientific reputation. Moreover, on a controversial subject, societal concerns could overshadow the science and with both in EFSA, the science would be at risk of getting lost. If anything, that could damage EFSA's credibility rather than enhance it.

Secondly, a 'concern assessment' is already available to risk managers on ethical issues. We think they should make more use of it and perhaps set up a group of experts on whom they could call for a more in-depth study of concerns when they are faced with framing new laws. Alternatively, increased pre-regulation consultations and stakeholder dialogue might be the way forward – and we are seeing parts of the Commission doing more of this already. We think it would be confusing to have the European Group on Ethics in Science and New Technologies available to be called by risk managers but the concern assessment buried in EFSA. Concern assessment is more closely related to management tools like impact assessments that are also in the remit of the Commission. Essentially, what a concern assessment will help managers to do is more effective 'change management' (to use business speak) when they introduce new laws. This is clearly a risk management issue because it will differ by culture.

Thirdly, EFSA is stretched for resources to carry out its existing tasks let alone some associated ones which are in its original mandate, such as those concerning nutrition. We do not think it would be right to divert EFSA resources to concern assessment. Nor do we think EFSA currently has any expertise on this. Moreover, there is also a risk that if EFSA were to do concern assessment it would lead to delays in scientific risk assessment. We do think more resources should be applied to concern assessment, however, but as part of the risk managers' toolbox.

Fourthly, EFSA was set up under the Regulation 178 of 2002 to provide scientific advice and technical support on food and feed safety. Changing its remit when it has been fully functioning for only a couple of years – given the issues around its location – seems premature and an unnecessary distraction to the developing core of its work.

12.4.8 Other Issues

Once the concern assessment is removed from EFSA and put in its rightful place in the risk managers' toolbox, then it seems that some of the other structural innovations suggested are no longer necessary; and that it may be more of a question of improving existing procedures of consultation from risk managers, for example, then putting in a new structure or new system. As for whether a precautionary approach is taken to management measures, this seems to us entirely for risk managers to deal with, based on the input they receive from scientific risk assessment, societal risk assessment, and stakeholder consultation.

The researchers focus, at some length, on different purposes of participation at different governance stages. We agree there is more scope for participation, particularly in terms of stakeholder dialogue, most particularly in the area of societal concerns. We think the best way of tackling this, however, is through risk managers specifically asking for a concern assessment from experts on societal concerns, and using more and earlier consultation techniques and stakeholder dialogue techniques to gather input. The internet is a good tool for that: timely consultation as a text is being drafted is then quite possible. It is not clear that the rather complex set of participation structures proposed would bring more clarity to the system nor enable all those who wished to participate to do so. Moreover, the elaborate structures proposed are partly there because a concern assessment is put in EFSA – once it is not in EFSA then the need for such widespread participation between stakeholders, risk managers and EFSA all together becomes lessened. It is the risk managers who need to interact with the stakeholders: ultimately they have the accountability to draft laws that will work in the political culture concerned.

The idea of an interface advisory committee which could adopt advisory opinions on terms of reference and on evaluation of cases and even on bundling of cases was not convincing. Different interest groups will be interested in different cases. Whereas risk assessors and risk managers have clear decision-making rights and responsibilities, it was not clear that other players in such a committee would have any responsibilities to balance against their right to give a view on a particular issue. There would be major issues about who should be represented on such a committee and concern about it slowing down the process of discussion and decision-making between the risk assessor and risk manager. The method of internet consultation on particular issues where respondents have to state who they are, seems preferable to this, leaving the responsibility for managing the process of gathering input clearly with the risk managers.

12.4.9 Conclusion

Against the six principles set out earlier, the proposals from the WP5 team do provoke a good discussion. However, the splitting of the scientific risk assessment into three categories ('risk-based', 'precaution', and 'prevention') was not ultimately convincing, while the fourth category, the examination of societal risk, is of such a different order that it does not seem to belong in EFSA. That risk managers should pay more attention to societal risk factors, however, is not in doubt. This should be done by ensuring that risk managers do have a panel of experts whom they can consult and that risk managers take timely measures, as regulations are being developed, to hold early stakeholder consultations using the internet to ensure that they are aware of views and expertise outside of the institutions. As far as more formalized structures of involving stakeholders in the process were concerned, the risks of confusing the accountability of risk assessors and risk managers for their decisions and the risks to the efficiency and timeliness of the process seemed to outweigh the benefits. Most of these benefits seem likely to be obtained by timely consultation over the internet. In short, by instituting better processes for consultation it would seem much less necessary to introduce new structures.

I would like to thank the SAFE FOODS WP5 team for involving me in their hard work and discussions that have proved so stimulating on such an important subject as food safety governance.

Glossary

Ambiguity

A state of knowledge under which incomplete information or divergent informed understandings preclude full confidence in the bounding, partitioning, characterising or prioritising of the possible *outcomes*.

Assessment

The process of gathering relevant information for the purpose of informing decision making concerning the relative merits and drawbacks of a range of different possible decision *options*.

Certainty

A state of knowledge under which there exists no *incertitude*. In other words, knowledge is judged to be definitive and complete concerning both the nature and the eventuation of the *outcome* in question.

Concern Assessment

A systematic, scientific process of gathering and analysing data on social responses to *threats*, insights on risk perception, and information on other specific 'secondary outcomes'.

Dose

The magnitude of *exposure* to a potentially *hazardous* agent or property.

Dose–Response Assessment

A step in *risk assessment* involving the determination of the magnitudes of the causal relationships between the *dose* and the *response*.

Evaluation

The process of determining the value-based components of making a judgement on a given *threat*, as informed by *assessment* and as necessary for *management.*

Exposure

The magnitude, *likelihood* or frequency of contact between a (human or environmental) system of interest and a potentially *hazardous* agent or process.

Exposure Assessment

A step in *risk assessment* involving determination of qualitative forms or quantitative magnitudes of possible types of contact between human or environmental systems and potentially *hazardous* agents or processes.

Food Safety Governance

Includes, but also extends beyond, the three conventionally recognised elements of *risk analysis* – risk assessment, risk management, and risk communication. It comprises matters of institutional design, technical methodology, administrative consultation, legislative procedure, and political accountability on the part of public bodies, and social or corporate responsibility on the part of private enterprises. It also includes more general provision on the part of government, commercial and civil society actors for building and using scientific knowledge, for fostering innovation and technical competences, for developing and refining competitive strategies, and for promoting social and organisational learning.

Flexibility

A property of an individual decision *option* relating to the degree to which this is subject to deliberate intervention in order to effect structural or functional change in the face of changing circumstances.

Food Safety Communication

The process of two-way communication with *stakeholders* and the wider public in order to frame, inform and convey the rationale and outcomes of *assessment*, *evaluation* and *management*.

Framing

Relates to 'risk assessment policy' (in the terminology adopted by Codex Alimentarius) and is made up of three activities – '*review*' of the technical and institutional conditions relating to food safety in its broadest sense; '*referral*' of specific *threats* to the *assessment* authority for the process of *screening*; and the setting of '*terms of reference*', upon which the *assessment* authority will base the *assessment*.

Hazard

A possible source of harm to human beings or the environment.

Hazard Characterisation

A stage in *risk assessment* involving the qualitative and/or quantitative evaluation of the possible magnitudes of *hazards*.

Hazard Identification

A step in *risk assessment* involving the determination of biological, chemical, and physical agents or properties capable of causing adverse health or environmental effects.

Ignorance

A state of knowledge under which there exists both *uncertainty* about *probabilities* and *ambiguity* over possible *outcomes*. In particular, ignorance involves exposure to the possibility of surprise.

Incertitude

A term used in a precise and specific fashion to refer collectively to real-world combinations of states of *risk*, *uncertainty*, *ambiguity* and *ignorance*.

Indeterminacy

A particular set of conditions contributing to a state of *ignorance*, under which relevant causal processes of the phenomena in question are open, dynamic, recursively linked to the observer or otherwise incompletely understood.

Interface Institution

A collective term that refers to the innovative mechanisms allowing communication and co-ordination between *assessment* and *management* activities (specifically the '*Internet Forum*' and the '*Interface Committee*' in its two variants). In this regard, the term 'institution' is used in a broad sense and does not relate to the formal EU institutions of the European Parliament, Council, Commission and Court of Justice.

Interface Committee

A food safety governance committee made up of assessors, managers and stakeholders that serves to act as an interface between *assessment* and *management* governance stages. The two variants of such a committee highlighted in this book are named the 'Interface Advisory Committee' and the 'Interface Steering Committee'.

Internet Forum

The basic recommendation for creating a food safety interface structure; it is an online function which should act as a site for information dissemination and exchange of views associated with every stage in the governance process.

Intrinsic property

A quality that is intrinsic to a potentially *hazardous* agent or process and is of relevance in the *assessment* of the agent, but which is not necessarily of itself in any way *hazardous*.

Irreversibility

An *intrinsic property* of a potentially *hazardous* agent or process or its derivatives arising where one or more of the consequences of its use are not readily subject to restoration to the state preceding this use.

Likelihood

The frequency or plausibility of the chance that a defined outcome will in fact eventuate. Where this is expressed in quantitative terms, it is a *probability*.

Management

A term used to refer to the process informed by *assessment* of decision making, implementation of measures, and monitoring of how these measures perform in practice.

Option

A particular possible course of action that may be adopted in decision or policy making, either individually or as part of a *portfolio*.

Outcome

The consequences of a particular course of action or state of the world.

Persistence

An *intrinsic property* of a potentially *hazardous* agent or process or its derivatives arising from the propensity to be retained in the environment in an active form over long periods of time.

Portfolio

A mix of different decision *options* pursued concurrently.

Precaution

An approach to *assessment* and *management* prompted by the *precautionary principle*, under which deliberate attention is afforded as much to *uncertainty*, *ambiguity* and *ignorance* as to the narrower condition of *risk*.

Precautionary Assessment

The use of a wide variety of broad-based approaches at the earliest stages in an innovation or policy making process, extending beyond conventional quantitative, expert-based techniques of *risk assessment*.

Precautionary Principle

A legal and policy principle adopted in various forms under many national and international instruments, which holds important implications for the conduct of *assessment* and decision making under *uncertainty*.

Presumption of Prevention

The appropriate response to a certainly and unambiguously serious threat, in which *assessment* is bypassed and preventative *management* measures are prioritised.

Probability

A quantitative expression of the *likelihood* of some defined *outcome* in terms of a numerical value between 0 and 1, where 0 indicates impossibility and 1 indicates *certainty*.

Resilience

A property of a *portfolio* (or individual decision *option*) relating to the capability of sustaining functional value despite short term episodic shocks arising in the external environment.

Response

The severity and/or frequency of adverse environmental or health effects associated with an *exposure* to a potentially *hazardous* agent or property.

Risk

A state of knowledge under which the range of possible *outcomes* has been well characterised and there exists sufficient information confidently to determine the *probabilities* associated with these outcomes.

Risk Analysis

A term used (especially in the USA) to refer to the entire process of *hazard identification*, *risk assessment*, *risk management* and *risk communication*.

Risk Assessment

A range of *assessment* techniques involving systematic characterisation of *likelihoods* and *outcomes* (usually through the determination of *probabilities*) in order to inform the prioritising of different decision *options*.

Risk Characterisation

A step in *risk assessment* involving the collection and analysis of all relevant evidence deemed necessary for informed decision making on the tolerability or acceptability of a particular *risk*.

Robustness

A property of a *portfolio* (or individual decision *option*) relating to the capability of sustaining functional value despite long term enduring change in circumstances.

Screening

Involves the preliminary characterisation of the *threat* in question in order to select the most appropriate form(s) of *assessment*.

Stakeholders

The full range of social actors who stand to be affected by decision making or who perceive themselves to hold an interest in its *outcome*.

Threat

A term that may be used in a general sense such as to include reference to both *hazard* and *risk* depending on the context.

Transparency

A quality and principle of good governance such that the natures of motivating reasons and priorities, analytic–deliberative processes and *outcomes* are readily accessible to detailed scrutiny by stakeholders.

Ubiquity

An *intrinsic property* of a potentially *hazardous* agent or process or its derivatives arising from the quality of being widely distributed in space, across ecological systems, or throughout different environmental media.

Uncertainty

A state of knowledge under which the range of possible *outcomes* has been well characterised, but there exists insufficient information confidently to determine the *probabilities* associated with these outcomes.

Vulnerability

A propensity on the part of environmental or human systems, ecological taxa or social groups of being exposed to possible harm from a potentially *hazardous* agent or process.

Index